Microfungi

Microfungi

Suzanne Gravesen
Jens C. Frisvad
Robert A. Samson

Munksgaard

Microfungi

1st edition, 1st printing 1994

Copyright © 1994 Suzanne Gravesen, Jens C. Frisvad,
Robert A. Samson and Munksgaard

Composition and cover by Kirsten Krog
Typesetting and reproduction by
HighTech PrePress A/S, Copenhagen
Printed in Denmark by
Special-Trykkeriet Viborg a/s

ISBN 87-16-11436-1

Cover: ground pepper flora (see page 16).

Contents

Acknowledgements

We would like to express our sincerest gratitude to Margaret Flannigan, who patiently edited the text in our very generous stream of manuscripts. We also want to thank our friends and colleagues for providing illustrations from their areas of expertise. Jane Piasecki provided invaluable practical help and support with the manuscript. We also gratefully acknowledge fruitful discussions and comments on the first drafts of the book from Ole Jørgensen. Ellen Kirstine Lyhne, Marjan Vermaas and Karin Vermeulen are thanked for their skilful assistance.

Preface

Our intention with this book has been to give a presentation of the filamentous fungi – the moulds – by gathering together information from many different disciplines concerning these fungi, which otherwise would have to be collected from a variety of sources, if at all available.

We have emphasized the most common moulds in daily life, and thus this book does not deal with fungi occurring as diseases on plants and animals or macrofungi such as the mushrooms. Perhaps the engineer will focus on information on destruction of building material and moisture requirements for mould growth in houses, while the chemist working with food technology will benefit from the information on specific metabolites produced by the individual mould species.

The many serious health effects of these metabolites will form valuable knowledge for various groups of people working in the area of health and medicine.

We hope that the text, as well as the illustrations, will serve as an inspiration and an aid in mycological teaching as well as a useful tool for everyone with an interest in the fascinating world of microfungi.

Chapter 1

What is a fungus?

In nature the kingdom of fungi is a very special assemblage of organisms.

The fungi neither belong to the plant nor to the animal kingdom. They differ from plants as they lack chlorophyll and the subsequent ability to perform photosynthesis. They can be distinguished from animals because they have no organs for food uptake; no mouth or gut.

The obvious difference between mammalian and fungal cells is that fungal cells are encased in a carbohydrate-containing wall. Most fungi consist of cells which are oblong and sausage shaped or like filaments. These cells, called hyphae, collectively form a mycelium.

The fungi may be subdivided into different (artificial) groups such as the yeasts, the mushrooms (fleshy fungi) and the organisms emphasised in this book, the filamentous microfungi (the moulds). Reproduction is either sexual or asexual. Most of the fungi described in this book

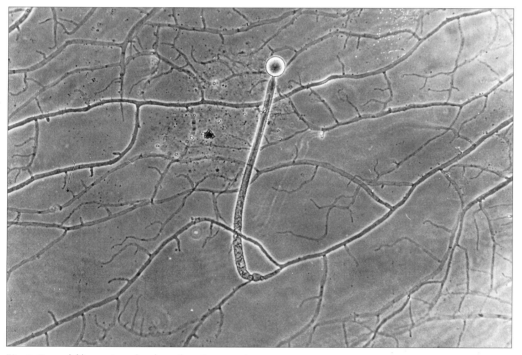

Fig. 1. Fungal filaments or hyphae, forming a web called the mycelium. The pinhead-like structure is the sporangium. The spores liberated from this will germinate and develop into hyphae forming the new mycelium. The fungus depicted is a *Mucor* species.

reproduce themselves without any previous meeting between a "male" and "female" mycelium.

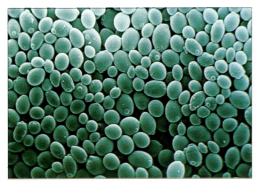

Fig. 2. SEM of typical cells of the beer yeast (*Saccharomyces cerevisiae*), which reproduce themselves by budding.

The professional terms for the propagules are spores when dealing with sexually reproducing fungi and conidia from the asexual organisms, but very often the two terms are interchangeable.

Fig. 3. *Amanita caesaria* one of the few edible species of this genus. It is a mushroom belonging to the Basidiomycetes. Most other species of *Amanita* are highly toxic.

Most fungal spores are 2-20 μm in size and of different shapes and colours. They are highly adapted for survival and dispersal. The cell-wall protects against desiccation and is often pigmented with e.g. melanin, making the fungus less vulnerable to radiation damage from ultraviolet light from the sun. Fungal-

protecting melanin is of a different chemical composition from that found in animal cells.

Fungi may also produce other forms of survival structures which are often very resistant to mechanical, chemical or biological attack; these are chlamydospores and hard stone-like sclerotia.

The spore-bearing structures will develop from the mycelium under certain conditions. In the microfungi these structures are called sporangiophores or conidiophores. They serve as reproductive organs as they may form millions of spores/conidia, which are liberated into the air – in nature from soil and plants – and mainly dispersed by the wind.

Maximum wind pick-up and dispersal occurs in dry weather. Spores with dew-dispersal or splash-dispersal will be scattered to the greatest extent in damp or rainy weather.

After liberation and dispersal, the fungal spores may germinate outdoors as well as indoors where suitable conditions for growth are present.

Microfungal spores will be found almost everywhere but they will only actually germinate and grow if organic material is present and the moisture content is above a certain level.

Fig. 4. Conidia, asexual spores of *Aspergillus oryzae* produced in long dry chains.

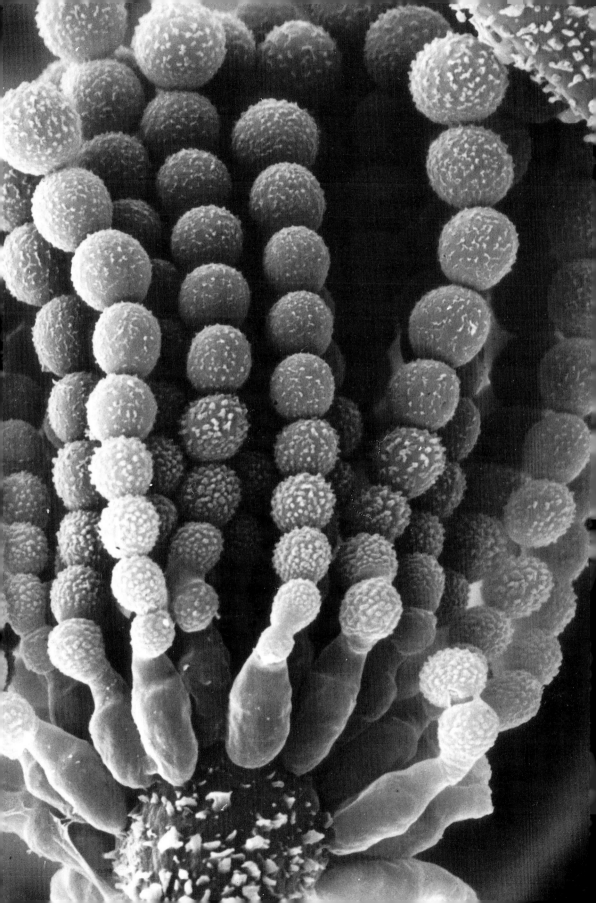

Fig. 5. A forest of conidiophores with conidia of *Penicillium glabrum*. Note the conidia at the bottom already liberated from the conidiophores.

In nature, plant and animal debris will be degraded to their simplest chemical components by means of fungi. Together with the soil bacteria the fungi are the most important organisms in connection with the degradation of biological material. We could name them "garbage-men of nature".

At least one fungal species will always be able to utilize any organic material as a nutrient source: e.g. hay, crops, food-stuffs, animals and their droppings, wood, paper, linoleum, paint, to mention just a few – even fuel oil in aircraft may serve as a substrate for fungal growth provided sufficient moisture is available.

Fungi also serve as food for insects and as symbiotic partners with algae in lichens.

These diversified abilities form the basis for their use and industrial exploit-ation, e.g. in the food-processing indus-try and the pharmaceutical/medical industry.

The same characteristics may, how-ever, be the cause of the damaging or toxic effects of the moulds, as will be seen from the chapters on biodeteriora-tion and toxigenic moulds.

Apart from their harmful effects, it is important to realize that humanity is highly dependent on these "garbage-men of nature".

Without their activity, the recycling of organic material would cease, with serious consequences for the environ-ment and for all human beings.

Fig. 6. *Mycena galericulata* infected with a mucoraceous fungus. Also mushrooms may be substrates for microfungi in their decaying activity.

Naming of a fungus

Fungi are named according to the so-called binomial system of nomenclature using the name of the genus and the species, e.g. *Penicillium chrysogenum.* This nomenclature is also applied to plants and animals, and the names of the fungi are set down in the International Code of Botanical Nomenclature. When a researcher has discovered a new fungus, he or she can describe it according to certain rules and the species will carry his or her name, e.g. *Penicillium chrysogenum* Thom.

If in later studies another species, described earlier, proves to be identical, then the older name has priority. For example, *P. notatum* (1911) Westling was described one year later than *P. chrysogenum* (1910) and is thus a synonym of the latter.

In many cases a species incorrectly placed in a genus can be put in another genus and a new combination is made, e.g. *Eurotium oryzae* Ahlburg, which is now *Aspergillus oryzae* (Ahlburg) Cohn. Many fungi have a life cycle in which asexual and sexual stages occur. These stages can bear different names, but the correct name should represent the whole fungus and this is called after the sexual stage. An example is *Emericella nidulans,* where the name represents both the ascomycete sexual stage (also called teleomorph) and the conidial stage (or the anamorph). The latter, however is also classified as *Aspergillus.* Although *Emericella* is the correct name for this fungus, in the literature the name *Aspergillus* often occurs. Classification of fungi is still in a state of flux and names are regularly changing. Some fungal names cannot be used anymore because they were used for several species by different authors or they were too poorly or ambiguously described.

Chapter 2
Biodeterioration – spoilage moulds

Mould spores are present almost everywhere, and a constantly elevated humidity in a given material will inevitably lead to microbial growth and subsequent damage to products such as textiles, leather, wood, foodstuff, paper, paint and many other things of organic origin. Even fuel oil in aircraft may be attacked by a mould (*Amorphotheca resinae*).

Today more than 6000 fungal genera are known. These genera comprise at least 70,000 species and even mycologists are not able to establish a comprehensive knowledge of all the fungi.

However, handling of practical problems with spoilage moulds depends in part on such knowledge, as these moulds have very different characteristics and very specific demands for growth in different environments.

For this reason, only a limited number of fungi are relevant in connection with the different materials, but often a number of species occur within an area.

The more sophisticated tools for fungal identification which have been elaborated in recent years have revealed a more precise picture of a specific spoilage mould flora related to the different types of biotope, which could be foodstuffs or building materials.

Fig. 7. Able to make mycelial clumps during growth in engines of aircraft by utilising the carbon source in the fuel oil, *Amorphotheca resinae* may create technical problems. In the picture conidiophores and conidia are seen.

Spoilage of food

Foodstuffs infected by moulds can be spoiled by production and subsequent dissemination of acids, compounds with an unpleasant taste, toxins and many other metabolic products from the fungi.

Fig. 8. Ground pepper spread on a Petri-dish containing the substrate dichloran rose bengal chloramphenicol (DRBC-agar). DRBC is a good selective medium for culturable fungi as the dichloran and rose bengal prevent overgrowth by the fast growing fungi, and chloramphenicol prevents bacterial growth. Note that the ground pepper is not infected, but the conidia present in the pepper have a potential for growth, given favourable conditions.

16

Around 50 fungal genera are relevant as spoilage organisms, and the most important among these are species of *Aspergillus*, *Penicillium* and *Fusarium*. Although many of the 50 genera will be present on the different foods, their germination and subsequent growth is dependent on the product itself as well as the environmental factors such as water activity, pH and temperature.

Contamination by moulds in cereals and fruits may cause loss of nutritive value, loss of energy, loss of ability to germinate, and, as well as the above mentioned factors: discolouring, change of taste, unpleasant smell (musty odour) and formation of toxins and antibiotics.

Relationship between mould growth and deterioration of quality in food is often very difficult to establish. The moulds – (mycelium and/or spores) – are not always visible or detectable at the stage when the problems are present.

Fig. 9. Oranges are often spoiled by *Penicillium digitatum* and *P. italicum*. During growth a characteristic smell from the volatiles produced can be observed. In this case the spoilage is caused by *P. digitatum*.

Fig. 10. Grapes infected with *Botrytis cinerea*. The fungus may cause serious problems as a spoilage agent, principally of soft fruit and vegetables. In wine production the infection is, however, positively utilised. The so-called noble-rot which it causes in grapes, in French *la pourriture noble* and in German *Edelfäule,* gives greater taste and sweetness to the wine.

17

Growth of *Penicillium expansum* in apples used for apple juice causes the formation of the mycotoxin patulin and development of the enzyme pectinase, with subsequent destruction of the apple juice, which changes the taste of the juice. During the production process, which includes pasteurization and filtration of the juice, the mould itself is killed, but being heat-resistant, the unwanted metabolic products may still be present.

In spite of heavy dilution during the processing of different foodstuffs, components with an unpleasant flavour may still remain and create quality problems in the final product, as even minute concentrations may have a strong spoilage effect on the taste. Enzymes from *Penicillium commune* and *P. roqueforti* may degrade the common preservative sorbic acid to 1,3-pentadiene and cause a unpleasant sour smell in the product.

Another example of spoilage from fungal components is the production of aflatoxins, specific toxic metabolites produced under special circumstances by *Aspergillus flavus*, *A. parasiticus* and *A. nomius* that infect nuts and other crops.

Apart from acute toxic effects after ingestion of highly infected food, long-term toxic effects have revealed aflatoxin B_1 to be the most carcinogenic naturally occurring substance ever encountered. The toxicity of this metabolite is so high that a basic threshold limit value for the amount provoking chronic toxic effects is difficult or impossible to establish.

Unpleasant odours may also create problems. The compound 2,4,6-trichloroanisol has a threshold value of 0.00003 microgram/kg. An important fungicide used for cardboard boxes and paper bags, pentachlorophenol, often contains impurities of 2,4,6,-trichlorophenol, and these substances are transformed by certain *Penicillium* and *Aspergillus* species by methylation to the strong-smelling anisol compound mentioned above. The unpleasant odour is extremely obvious in the end product (e.g. dried fruit and cocoa exported from Australia to Europe) even when the fungus is not detectable.

Fig. 11. Biodeterioration of rye bread caused by *Penicillium roqueforti*. Rye bread contains naturally occurring acetic acid and in some cases propionic acid, which normally prevent mould growth. However, *P. roqueforti* is resistant to these acids.

Table 1.
Important moulds in food decay

Fungal species	Attacked foodstuff
Aspergillus flavus	nuts, acid-preserved crops, corn
Botrytis cinerea	grapes, soft fruits
Endomyces fibuliger	gas-packed bread
Eurotium herbariorum (*Aspergillus glaucus*)	marmalade, jam
Fusarium moniliforme	corn
Penicillium aurantiogriseum	cereals
P. verrucosum	cereals
P. viridicatum	cereals
P. digitatum	citrus fruits
P. expansum	pomaceous fruits, nuts
P. italicum	citrus fruits
P. roqueforti	rye bread
Rhizopus stolonifer	strawberries
Yeasts	beverages and liquefied foods

Damaging effects on building materials

The presence and growth of mould spores indoors may lead to health problems of different severity as well as damage to the attacked materials.

Since moulds possess a variety of different very potent enzymes and acids, they can destroy organic materials very efficiently.

During growth of the moulds, production and excretion of these enzymes may cause the material to disintegrate. This, in combination with production of discolouring dyes, is often seen on mouldy wall paper, wood or other material, which contains cellulose attacked by microfungi. Blue-stained wood is a result of such an attack. In certain periods of time, this discoloration was used positively for decorative purposes.

Mould spores will be introduced to the indoor environment, e.g. through open windows during summer and autumn, brought in by dirty footwear or simply attached to dust particles. They can grow on foodstuffs, on flowers or other plants or they can grow on different types of building material.

Fig. 13. Mouldy wallpaper in a flat with insufficient insulation. Room temperature, a high water activity of the material and optimal nutrition from the wallpaper adhesive will make the moulds flourish, as seen here.

Fig. 14. Heavy mould growth in a prefabricated house where a builder wrongly installed the vapour barrier on the cold side of the plywood sheeting, causing considerable moisture damage, mould growth and infestation with house dust mites. As a result, many of the families inhabiting the houses suffered from allergies and other adverse health effects.

Fig. 12. Paper stored in a humid basement with subsequent attack of cellulose decomposing microfungi causing disintegration and discolouring of the fibres.

20

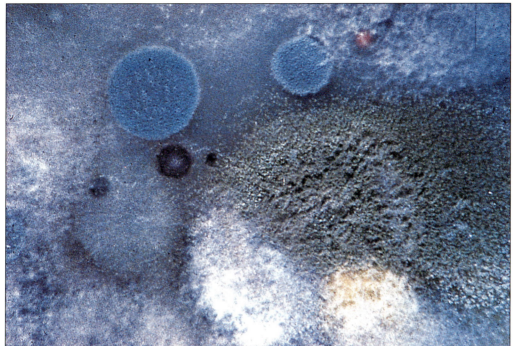

Fig. 15. Culturable moulds from dust collected from a 10-year-old carpet in a school and cultivated on V-8-agar. The viable mould spores comprise only about 1% of the total number of spores present.

Fig. 16. Water-damaged gypsum board infected with *Stachybotrys chartarum*, a highly toxic mould. A growing number of cases with *S. chartarum* causing indoor climate problems are reported.

The key to fungal growth in buildings is a moisture supply. If the mould spores find suitable conditions, which could be moist basements, walls, ceilings or window frames with a high water activity (a_w), they will grow and damage the material infected.

The assumption that a relative humidity (RH) of less than 70% will secure no growth of mould is a myth. The water activity (a_w) of the building materials will be the determining factor. Water activity a_w is the expression of the available moisture at equilibrium and is defined as:

$$\frac{\text{vapour pressure of water in substrate}}{\text{vapour pressure of pure water}}$$

Optimal growth conditions for the mesophilic moulds are an a_w in the actual substrate ranging from 0.95 to 0.99. For the yeasts: 0.88 to 0.99 and for the xerophilic moulds 0.65 to 0.90.

Examples of first invading moulds (primary colonizers) are *Penicillium chrysogenum, P. brevicompactum, P. commune, P. expansum* and *Aspergillus versicolor*. Secondary colonizers are *Cladosporium herbarum, C. cladosporioides* and *C. sphaerospermum*. Tertiary colonizers are *Ulocladium sp., Fusarium moniliforme, Phoma herbarum* and *Stachybotrys chartarum*. Numerous reports on building materials damaged by *S. chartarum* have been published, including descriptions of health problems.

Old, dirty, wall-to-wall carpets placed in service buildings such as schools, kindergartens and offices offer an excellent substrate for growth of deposited mould spores if the carpets are supplied with adequate moisture, either due to accidental water damage e.g. in winter when people will walk in with wet shoes or boots, or on purpose.

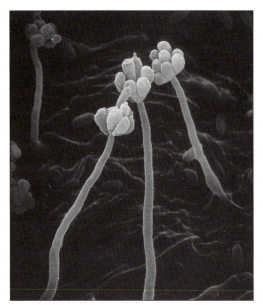

Fig. 17. SEM of *Stachybotrys chartarum* showing conidiophores and conidia.

Fig. 18. *Serpula lacrymans*, the dry rot fungus, a macrofungus, mentioned here as it has great potential as a deteriogen of building material. Insurance companies usually cover damage caused by *S. lacrymans*, but not deterioration of materials by moulds.

Fig. 19. When found on building materials, *Penicillium chrysogenum* is an indicator organism for moisture problems.

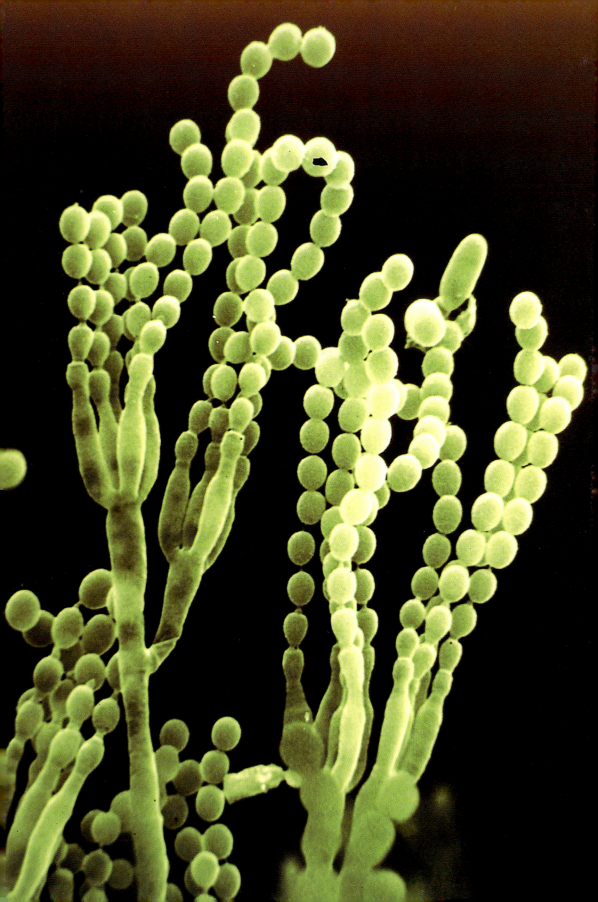

Table 2.
Important moulds in damp buildings

Fungal species	Where found
*Alternaria alternata**	moist window-sills, walls
Aspergillus versicolor	damp wood, e.g. floor construction
Aureobasidium pullulans	outlets from bathroom, kitchen
*Cladosporium herbarum**	moist window-sills, wood
*C. cladosporioides**	moist window-sills, wood
*C. sphaerospermum**	moist window-sills, wood
Eurotium (Aspergillus) repens	furniture, wallpaper
Penicillium brevicompactum	moist chip boards
P. chrysogenum	damp wallpaper, behind paint
P. expansum	moist chip boards, mineral wool
Rhodotorula spp.	moist basement walls
Stachybotrys chartarum	heavily wetted gypsum boards
Trichoderma viride	damp wood material

* These species may be part of the natural airborne microflora entering via the outdoor environment. Before concluding that mould contamination exists, the presence of areas growing moulds should be ascertained.

The latter could be the result of cleaning procedures which use shampoo and water or prevention of static electricity with humidification of carpets, as is sometimes recommended by authorities.

Microbial volatiles (mVOC)

Microbial volatiles are organic solvents such as alcohols, ketones, esters and hydrocarbons.

Mould growth in dwellings is often discovered because of the emission of characteristic musty or earthy odours produced by the moulds during growth. One of these heavy musty odours is 2-methyl-isoborneol produced by *Penicillium commune*. Other volatiles often encountered in damp houses are ethanol and acetaldehyde.

Frequently occurring *Penicillium* species indoors such as *P. chrysogenum, P. commune, P. glabrum, P. expansum* and *P. polonicum* produce 1-octen-3-ol, 1-pentanol, 3-octanol, 3-octanone 2-methyl-1-propanol and 3-methyl-1-butanol.

Another unpleasantly smelling volatile component is geosmin (1.10-dimethyl-trans-9-decalol) shown to be produced by *Chaetomium globosum* and by the actinomycete *Streptomyces*.

As is the case with the production of mycotoxins, water availability may have a considerable influence on volatile production by moulds. Experiments with growth of moulds on agar have led to the assumption that production of volatiles is inversely related to water activity (a_w) and that the volatiles are accumulated during stress periods in which they fulfil an osmoregulatory function, after which they can be released. However, it is also known that

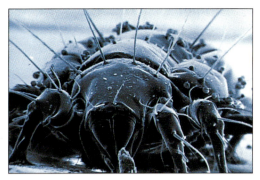

Fig. 20. SEM of a mite with fungal conidia on its body.

fungal volatiles have a strong ecological effect: e.g., 1-octen-3-ol is an attractant for certain mites and insects.

Most odours in damp buildings are caused by mould growth and in some cases by bacteria, especially actinomycetes but rarely by wood-rotting fungi.

Apart from their irritative effect, these microbial volatiles can be very troublesome.

Sweden, in particular, has had considerable mould problems because houses are built directly onto the soil above the underlying rock. The floor is then placed on a concrete slab, often without a waterproof barrier between the concrete and the soil.

Mould growth and unpleasant odours can also occur after leakage of pipes, presence of moisture in a sub-floor crawl space or wet attics.

Curtains, carpets and clothing of people from mouldy houses will absorb the unpleasant odours, and people coming from such mouldy houses very often carry these musty smells around, resulting in social isolation or teasing of their children in school.

Problems with microbial volatiles in water supplies may sometimes occur. These unpleasant odours can be caused by actinomycetes, but geosmin from

Chaetomium globosum may contribute along with a ketone a musty odour, produced by *Trichoderma* and *Aspergillus* (6-pentyl-o-pyrone), and benzyl cyanide may create a grassy odour produced by *Botrytis cinerea.* All of them have been isolated from bottom deposits of water supplies.

From field observations as well as from experimental research, there is growing evidence that microbial volatiles may have serious health implications in mouldy buildings.

In addition to the health implications, the economic consequences of microfungi attacking building material are of considerable importance.

Chapter 3
Moulds in biotechnology

The intention of this book – to introduce light and shade in the description of common microfungi, would not be fulfilled without a short summary of the different substances of commercial relevance produced by fungi.

Fungi are some of nature's most accomplished chemists, as they are able to synthesize very different compounds ranging from simple molecules such as ethanol or citric acid – examples of fungal metabolites – to complex, very special molecules such as aflatoxins and trichothecenes – more complex specific metabolites.

Microfungi used as chemical reagents

For people working in organic chemistry, it has been increasingly common to use fungal enzymes for several purposes. The enzymes are not isolated from the fungus, but the intact fungal cells with their specific enzymatic capacity are used for catalysing different processes.

The following examples will illustrate the useful role of moulds in chemical conversions.

A classical example is the use of baker's yeast, *Saccharomyces cerevisiae*

Fig. 21. Baker's yeast, *Saccharomyces cerevisiae*, still used in the oldest known fermentation processes: the production of beer, bread and wine.

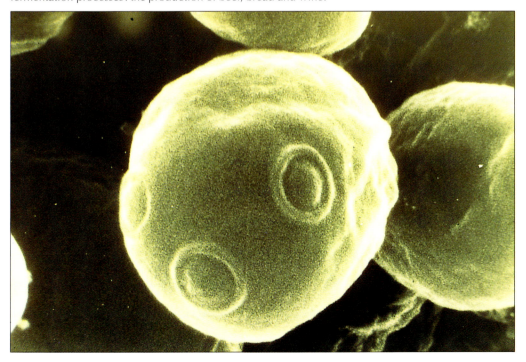

(=*diastaticus*), in the conversion of starch or sugar to alcohol.

Since the 1950s, filamentous fungi such as *Aspergillus niger, A. ochraceus* and *Rhizopus sp.* have been used in the pharmaceutical industry to introduce hydroxyl groups into the steroid skeleton of corticosteroids for production of preparations containing cortisone.

Fungal cells from *Aspergillus ochraceus, Fusarium solani* and *Rhizopus stolonifer* have been utilised for catalysing the production of mammalian metabolites used as experimental agents or for analytical standards.

Microfungi used for production of medicines

In the pharmaceutical industry, microfungi have played an important role either as producers of specific metabolites, such as penicillin (see below) or as producers of biomass from which important components are extracted either from the surface or from the interior of the cells.

Penicillin

In 1929 a microbiologist, Alexander Fleming, was working in his laboratory with cultures of the bacterium *Staphylococcus aureus* plated out on Petri-dishes. He noticed a contamination in some of his Petri-dishes with an airborne mould, *Penicillium chrysogenum* (formerly: *P. notatum*), which is occasionally a nuisance in the laboratory. At the same time he noticed a marked inhibition of the growth of the staphylococci.

This observation brought the most important and revolutionary contribution to modern medical treatment of many infectious diseases and started the new age of antibiotics.

The active component, the specific metabolite penicillin G – the first ß-lactam antibiotic – has the ability to inhibit the cell-wall synthesis of gram-positive bacteria. This ability has been further developed so that derivatives of penicillin can also combat gram-negative bacteria.

Fig. 22. A reconstruction of Alexander Fleming's observation. A culture of the bacterium *Staphylococcus aureus* has been contaminated with *Penicillium chrysogenum*. Penicillin diffusing out from the culture inhibits growth of the staphylococci. The genius of Alexander Fleming was not that he noticed this phenomenon but the conclusion he derived from his observation.

Cephalosporin

Another important group of specific metabolites with antibiotic activity are the cephalosporins derived from *Acremonium chrysogenum* (formerly *Cephalosporium*). They are closely related to the penicillins and are also able to inhibit cell-wall synthesis of gram-positive bacteria.

Cyclosporin

In the last decade, the most important therapeutic agent discovered has been cyclosporin A, a metabolite produced by the mould *Tolypocladium niveum*

(= *Beauveria nivea*). This substance has immunosuppressive activity, and this characteristic is used for modulation of the humoral response by inhibiting the production of the T-cell-growth factor: interleukin 2, an important mediator-component in the immune system.

Treatment with cyclosporins has dramatically improved the conditions for transplant surgery by reducing graft-rejection. As a result of this reduced morbidity, increased survival rates in transplant patients have been demonstrated.

Ergot alkaloids

The gap between toxic and therapeutic effects may be relatively limited. An example of this is the production of metabolites from the microfungus *Claviceps purpurea*, which as far back as the Middle Ages caused outbreaks of endemic poisonings following ingestion of infected crops. Abortion in both humans and cattle is also described. The causal agents – the ergot alkaloid derivatives, which have vasoconstrictive activity – have today found a therapeutic application in the treatment of migraine and in the control of motor activity of the uterus. Furthermore, they are used for therapy of Parkinson's disease and senile dementia.

Fig. 23. Disabled people painted by the medieval painter Pieter Breughel were casualties after eating poisened bread. At that time flour was often contaminated with sclerotia of ergot of rye, *Claviceps purpurea*. The fungus produces the toxic alkaloid ergotamine, which has a constrictive effect on the blood vessels, and this results in gangrenous extremities. Nowadays, ergotamine is used as a valuable drug in the treatment of migraine.

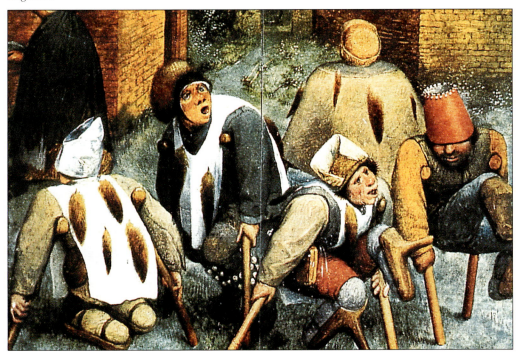

Mevinolin and asperlicin

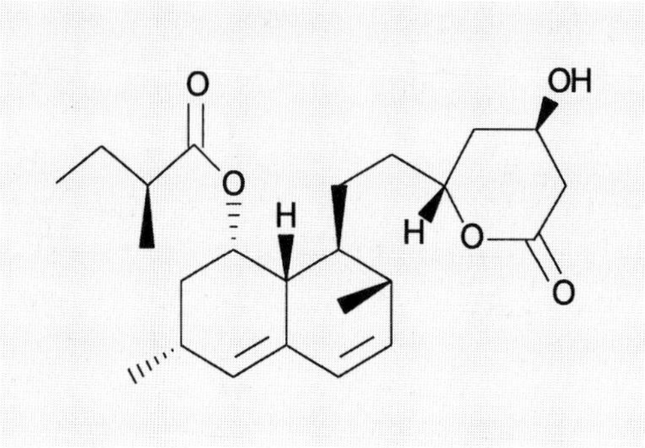

Fig. 24. Chemical structure of the specific metabolite mevinolin, produced by *Aspergillus terreus*. This agent has hypocholesterolaemic properties and is used in the treatment of hypertension.

Aspergillus terreus produces mevinolin, a metabolite which is a very potent inhibitor of cholesterol biosynthesis. It has been used for the treatment of a large number of patients with hypertension, especially in the USA.

Another specific metabolite, asperlicin, is derived from *Petromyces alliaceus* (formerly: *Aspergillus alliaceus*) and is used in the treatment of gastric disorders.

Allergen extracts
Extracts of the mainly proteinaceous components of fungal spores are used as reagents for the in vivo and in vitro diagnosis of airway allergy (asthma and hay fever). Standardised extracts, i.e. extracts with a reproducible content and concentration, are now available from some of the most common allergenic moulds, e.g. *Alternaria alternata, Aspergillus fumigatus* and *Cladosporium herbarum.*

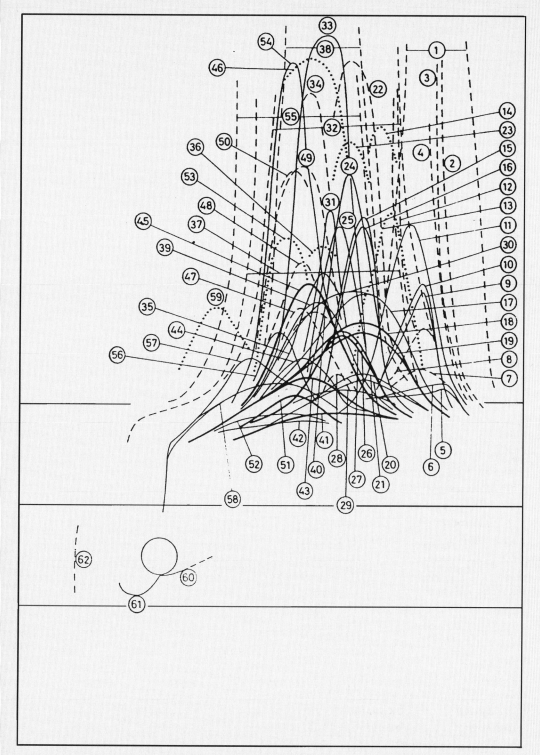

Fig. 25. Crossed immunoelectrophoresis (CIE) reference pattern of a crude *Cladosporium herbarum* extract. 10-15 of the detected antigens are allergenic in humans (Aukrust & Borch).

Fungal enzymes for food technology

The food-processing industry consumes more than half of the world's production of enzymes. Fungi are the primary producers of these enzymes, which are used to control many different features such as aroma, texture, colour and stability in different foodstuffs.

Proteases

Baking industry: in the USA a protease supplement from *Aspergillus oryzae* is added to two thirds of commercially produced bread to increase dough volume and to improve aroma, taste and the appearance of the crust and the internal texture, to mention just a few examples. *A. oryzae* is also a producer of α–amylase, which helps improve the dough.

Beer industry: modern beer brewing is also dependent on proteases from *A. oryzae* to ensure clarity of the beer and to assist in the masking process of the brewing; α–amylases produced by *Aspergillus niger* and *A. oryzae* are also used in this process.

Dairy industry: traditionally, rennet (the milk-clotting enzyme, a protease) was obtained from the stomachs of calves. Today more than one third of the worldwide production of cheese utilises microbial rennets. The most common fungi used for this purpose are *Rhizomucor miehei* and *Rhizomucor pusillus*.

Amylases and glucoamylases

Aspergillus niger (= *A. awamori*) and *Rhizopus oryzae* produce enzymes, amylases and glucoamylases, which are able to convert starch to different sugar derivatives used in food production.

Pectinases

Pectic substances are present in plants as structural components of the cell wall. Pectins are present during juice and fruit processing, and pectic enzymes derived from *Aspergillus niger*, *Rhizopus sp.*, *Penicillium chrysogenum* and *Botrytis cinerea* are used for degrading these substances.

In wine making, pectinases are used to remove different suspended particles and to improve the clarity of the final product.

Cellulases and hemicellulases

These are produced by e.g. *Trichoderma viride*, *T. reesei*, *Aspergillus niger* and different *Penicillium* species. They are used commercially in the production of beer, wine, food processing and food fermentation, e.g. to enhance the quality of tea and the flavour of mushrooms.

Fig. 26. SEM of zygospores (sexual spores) of *Rhizomucor miehei*. For many years the enzyme rennet, which is essential for cheese production, was obtained entirely from the stomachs of calves. However, for some time the ability of *R. miehei* to produce a protease enzyme with similar properties has been commercially utilised.

Lipases

Aspergillus niger, Rhizomucor miehei, R. pusillus, Rhizopus spp. and *Penicillium roqueforti* produce lipases used in the dairy industry for flavour development/modification, improvement of the ability of egg whites to be whipped and acceleration of cheese-ripening. The lipases are also now used for amending soap powder so that fat-containing spots can be removed from clothes etc. These enzymes are produced by *Aspergillus oryzae,* but the gene coding for the lipase originates from another (thermophilic) fungus *Thermomyces lanuginosus.* The latter fungus produces lipases which can work at high temperatures, which is often desirable during the washing of clothes.

Microfungi used in other industrial processes

Production of organic acids is highly dependent on certain microfungi. Organic acids are used in the food industry as well as in the production of paper, plastic and carpets.

Some important applications are described as follows:

Citric acid

At the beginning of this century, production of citric acid from *Penicillium glabrum* and later on from *Aspergillus niger* was known.

In terms of economics, citric acid is the most important organic acid. In 1987 consumption was valued at £300 million. The spectrum of application is rather broad, but the food industry is the primary consumer. It is used for tartness and pH control in jellies and jams and to adjust acid flavour in fruit juices, soft drinks and wine.

Citric acid is able to form complex bonds with heavy metals, and this ability is used for protection against oxidative deterioration in the flavour of foodstuffs.

Fig. 27. *Thermomyces lanuginosus,* a thermophilic fungus with enzymes capable of degrading lipids at 60°C. Genetic engineering has made it possible to express the gene in *Aspergillus oryzae,* which is easily cultivated.

Fig. 28. Conidial heads of *Aspergillus niger,* one of the most important industrial fungi, used especially for the production of citric acid.

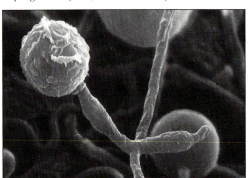

It helps to protect frozen food from loss of vitamin C and from enzymatic browning; furthermore, it is used as an antioxidant and for pH adjustment in cosmetic and pharmaceutical products. The strains of *Aspergillus niger* used are currently being improved to give higher yields.

Itaconic acid

Itaconic acid, produced by *Aspergillus terreus* and *Aspergillus itaconicus*, is used in the production of a co-polymer with synthetic resins and in synthetic fibre manufacture as an acrylonitrile co-polymer. It is also used in carpet backing and paper coating.

Another application is the control of plant growth, as the substance enhances root development.

Gluconic acid

This is produced by species of *Aspergillus* (mainly *niger*) and by *Penicillium*
purpurogenum and *chrysogenum*. It was used formerly to feed dairy cattle, but it is now only used to a limited extent in the food industry.

The primary function of gluconic acid is as a chelating agent. As it has the ability to function in alkaline solutions, even free caustic, it is used in cleaning and in metal-finishing operations.

Fumaric acid

This is produced by a number of *Rhizopus* species, and the most important commercial use is in the paper industry, where this acid improves the strength and stiffness of the paper.

Furthermore, it is used like citric acid in the production of fruit juices and gelatine desserts. A promising new area is its use as a food additive to increase feed efficiency in domestic animals.

Fig. 29. *Rhizopus stolonifer* with rhizoids (root-like structures) and sporangial heads.

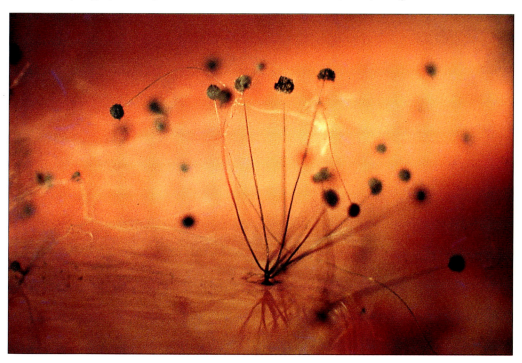

Food fermentation

Since ancient times fungi have been used for improving and protecting food-stuffs. Nowadays a number of food products and beverages are produced with the help of moulds and yeasts, particularly *Saccharomyces,* which is the essential fermenting agent for alcoholic beverages and bread. In the Western world, the fermentation of cheese by *Penicillium camemberti* and *P. roqueforti* and salami by *P. nalgiovense* is increasing. The useful role of moulds in foods in Oriental countries is even more evident. One of the most important species is *Aspergillus oryzae,* used for the fermentation of soya beans, resulting in soy sauce, a product which is widely used in Japan and South-East Asia and becoming very popular in the Western world too. Other species important for oriental food fermentation are *Rhizopus oligosporus* (tempe, a soya bean cake),

Neurospora intermedia (oncom, a peanut press-cake) and *Monascus ruber* (angkak, a red colouring dye produced from rice and used for beverages and other dishes). Some of the species used in the fermentation of food suitable for human consumption are now also tested for improving feedstuff for animals.

Fig. 30. A salami sausage fermented with *Penicillium nalgiovense.* The fungus is used partly for protection, partly to improve the taste, mainly in the Mediterranean countries and central and eastern Europe.

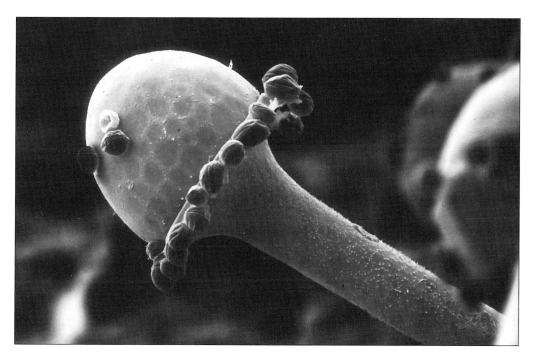

Fig. 31. SEM of *Rhizopus oligosporus*. This fungus has been used for centuries in the production of tempe, soybeans bound together with the mycelium of *R. oligosporus*.

Fig. 32. SEM of kefir organisms: *Candida kefyr,* other yeasts and bacteria belonging to the lactococci and lactobacilli. Kefir differs from yoghurt as it contains the yeast *Candida*. People in the Caucasus live longer, supposedly because of their daily intake of kefir.

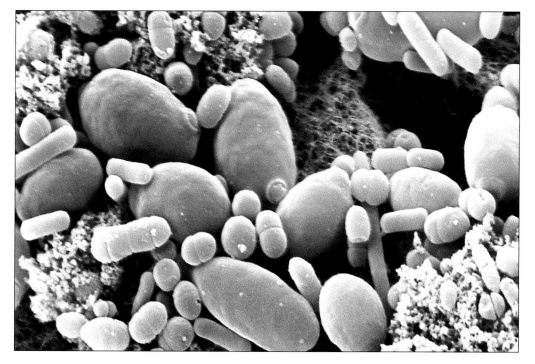

Fungal cells are very nutritious and contain fibres. Because the demand for meat as a protein source will be difficult to fulfil in the future, the possible use of several fungal species as a source of protein is under investigation. One of these products is quorn, a pure mycelial product of *Fusarium graminearum*. With added aromas or put into pasta or pies, the product seems to be quite popular in the United Kingdom.

Fig. 33. The role of fungi in food fermentation is particularly evident in tropical countries. Products are often improved by fermentation by using species of *Aspergillus, Mucor* and *Rhizopus*. In Indonesia tempe and oncom are common daily food ingredients, which are often produced in home industries using soybeans and the fungi *Rhizopus oligosporus* and *Neurospora intermedia*.

Fig. 34. Examples of various food products in which moulds and yeasts are important fermenting agents.

Fungi used in biological control

Insecticides, herbicides and helminthicides

Since uncontrolled use of pesticides has resulted in environmental pollution and development of resistant strains, research is now directed towards the development of agents which would act as more natural enemies.

An interesting but not completely successful attempt has been made to introduce spores of the nematode-killing fungus *Arthrobotrys* into the stomachs of cattle in order to protect them against the larval infestation of nematodes (roundworms), which inhabit the liver and reduce viability.

The idea is to kill the larvae while the cattle are in the cow-shed by means of different species of *Arthrobotrys*, which destroy the nematodes in their special fungal trap mechanisms. Among several species of *Arthrobotrys*, *A. oligospora, A. botryospora, A. pyriformis* and *A. vermicola* have been used.

The experiments also aim at preventing the nematodes from spreading to grassland where they can be taken up by the cattle again.

Fig. 35 (a + b). SEM showing how an infective larva of the worm *Cooperia oncophora*, which lives in the small intestine of cattle, is caught in a hyphal trap formed by the predaceous fungus *Arthrobotrys oligospora*. The fungus produces a sticky material, which efficiently secures the larva to the hyphae. After about 4 days the fungus will have digested the larva, leaving only the cuticle.

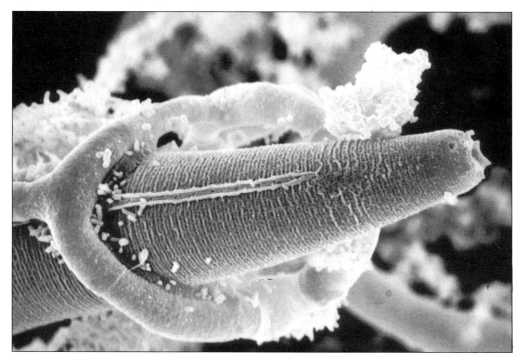

More successful attempts have been made with the fungus *Lagenidium giganteum,* whose asexual spores infect mosquito larvae and kill them.

Infectious spores of species of *Paecilomyces* have been used under field conditions against the voracious Colorado beetle, which destroys potato crops.

Results from the application of conidia from *Verticillium lecanii* in greenhouses against aphids seem to be promising.

Phytopthora palmivora has been marketed as a herbicide to protect citrus trees. By applying a liquid extract of the fungus to the soil around citrus trees, it is possible to kill seedlings and mature weeds.

There must be a large and yet undiscovered potential for the use of fungal components as biocides.

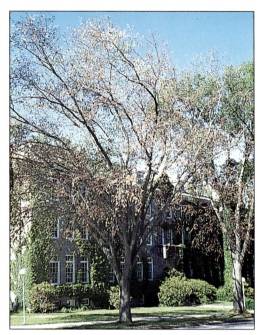

Fig. 37. A search is on for an effective biological control agent for the fungus *Ophiostoma ulmi* (*Ceratocystis*), which is rapidly killing most of the elm trees in Europe. The vector is the elm-bark beetle.

Fig. 36. *Beauveria bassiana,* a fungus named after Agostino Bassi, who discovered that silkworms in Italy were killed by this mould. The species has been used as biological control of several insect pests, including citrus weevils.

Fig. 38. Biological control of the citrus rust mite in Florida, USA. A suspension of spores or hyphae of *Hirsutella thompsonii* is sprayed on the fruits.

Fig. 39. Rhinoceros beetle killed by the fungus *Metarrhizium.* Strains of this fungus are used as potential biological control agents for various pests.

Fig. 40. A tropical moth species infected by the entomopathogenic fungus *Cordyceps tuberculata.*

Table 3.
Important moulds for industrial use

Fungal species	Product
Acremonium chrysogenum	cephalosporin
A. fusidioides	fusidin
Aspergillus niger	enzymes, citric acid
A. oryzae	enzymes, soya sauce
A. terreus	mevinolin
Cephalosporium acremonium	cephalosporin
Penicillium camemberti	fermented cheese
P. chrysogenum	penicillin
P. griseofulvum	griseofulvin
P. nalgiovense	fermented sausages
P. roqueforti	blue cheeses, flavour compounds
Rhizomucor spp.	enzymes, e.g. rennet
Rhizopus spp.	organic acids, enzymes
Tolypocladium inflatum	cyclosporin
Trichoderma reesei	cellulases

Chapter 4

Mycotoxins and mycotoxicoses

In nature, living organisms protect themselves, thereby protecting their genes, in many different ways. This can occur by colonising habitats through extremely quick growth, as is done by bacteria and some moulds; by emission of unpleasant smelling volatiles; by production of toxic substances; or even by attracting other organisms for effective dispersal.

Microfungi are able to produce a wide variety of different types of specific metabolite, e.g. "battle-substances", attractants and sex hormones.

Environmental factors

When growing in soil or any other environment, the microfungi very often produce toxins in order to protect a nutrient source against bacteria. This is easily understandable, because microfungi must protect their ecological niche in the soil, where they feed on plant litter and other organic debris in competition with the bacteria.

Specific metabolites produced by microfungi are toxic either to humans, vertebrates, insects, plants or other micro-organisms. The toxicological properties and degree of toxicity of these specific metabolites vary depending on the administration route, chemical constitution and concentration of these substances.

Specific metabolites as pharmaceutical drugs

Their capacity for producing such bat-

tle-substances is often retained during growth under laboratory conditions. This implies that some biologically active, specific metabolites from fungi can be used as pharmaceutical products, e.g. as antibiotics against pathogenic bacteria or fungi. Medication in connection with organ transplantations used to suppress the immune response to foreign

Fig. 41. A ball-and-stick model of the molecule meleagrin. This is another specific metabolite of *Penicillium chrysogenum*, with antibiotic properties. Light blue, carbon atoms; white, hydrogen; red, oxygen; and dark blue, nitrogen.

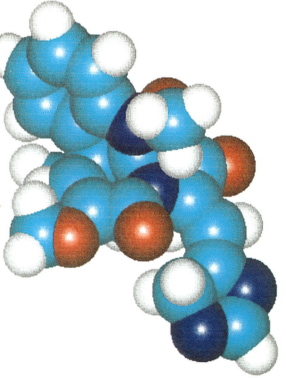

tissues and thus prevent transplant rejection has resulted in a considerable improvement with the use of fungal metabolites.

Although certain specific fungal metabolites produced in the laboratory have a very strong antibiotic effect, they may reveal toxic properties which make them unsuitable for use in human or animal medicine.

Well known examples of relatively nontoxic fungal-specific metabolites are penicillin and fucidin, used as very efficient drugs against certain bacterial infections. These are derived from a few species of *Penicillium* and *Acremonium*.

Griseofulvin, produced by *P. griseofulvum*, is toxic to other fungi, especially dermatophytes.

Another example of a specific fungal metabolite used for medication is mevinolin, which is used as a very effective cholesterol-lowering agent.

Fig. 42. Laboratory culture of *Penicillium griseofulvum*, a mould from which the drug griseofulvin, active against some fungal infections, is derived.

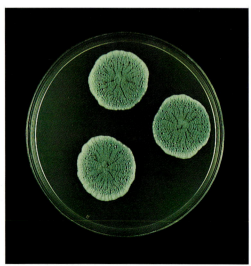

Mycotoxins – the toxic metabolites

When the toxic effects of the specific metabolites produced by a fungus exceed the useful antibiotic effects, the metabolites cannot be regarded as drugs and the word mycotoxin is used. This is defined in the following way:

"Mycotoxins are natural products (specific metabolites = secondary metabolites) from moulds which initiate a toxic response in vertebrates, when introduced in small concentrations by a natural orifice, i.e. the mouth, the respiratory system or the skin".

Hence the toxins produced by many mushroom species do not fit this definition, although these compounds can evoke a very strong toxic reaction.

Discovery of mycotoxicosis

One of the first documented diagnosis of mycotoxicosis was established in England in 1960, after an outbreak of a disease causing the death of more than 100,000 turkeys. A multidisciplinary team in London revealed that this unknown disease, which affected the liver and which at the same time occurred in ducklings, pigs and cattle, was due to ingestion of mouldy peanuts. These had been imported by ship from Brazil where they had been stored under unfavourable conditions. This had resulted in contamination of the nuts with a yellow-green mould: *Aspergillus flavus,* which subsequently produced a specific metabolite

Fig. 43. SEM of *Penicillium chrysogenum*, an important producer of the metabolite penicillin. From the early 1950s, the use of this first antibiotic drastically reduced the effects of bacterial infections.

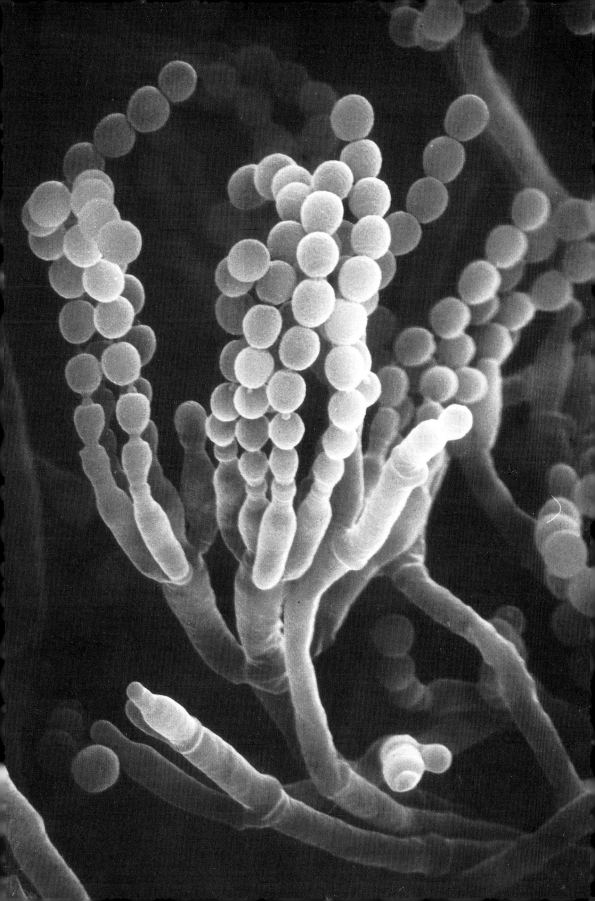

eventually named "aflatoxin" (from *A. flavus* toxin) in the nuts. Later research has shown that other toxins, such as cyclopiazonic acid from *A. flavus*, may also have been involved in the disease.

Effects on humans and animals

The toxic effects of ingested mouldy foodstuff are experienced particularly by experimental and domestic animals and are very different, comprising acute or chronic damage to the liver, kidneys, gastrointestinal tract, the heart, the central nervous system (CNS) and the immune system.

Fig. 44.

Chemical structures of the tremorgenic mycotoxin penitrem A (causing spasms of the muscles) and the neurotoxin roquefortine C. Both toxins are produced by *Penicillium crustosum*, mainly found on nuts and occasionally on cheese.

Four cases of dogs dying from penitrem A intoxication have been reported in different countries (Australia, Canada and USA). In all cases, different cheese products were heavily contaminated with *Penicillium crustosum*, which produces both roquefortine C and penitrem A. The latter toxin, causing strong spasms of the muscles (tremor), ap-

peared, however, to be the sole factor responsible.

In the Canadian case, roquefortine C – also produced by *P. roqueforti* and used for fermenting the particular blue cheese which the dog ate – was suspected of being the cause of the poisoning, but this seems highly unlikely. Recent research has shown that *P. roqueforti* and *P. crustosum* often occur together in such deteriorated products.

Another example of poisoning by these two organisms was the intoxication of a man who drank a beer containing a fungus "ball". During initial investigation, it was shown that the *P. roqueforti* recovered could produce roquefortine C and isofumigaclavine A, but after re-examination of the culture, *P. crustosum* as well as *P. roqueforti* were found, and the former was capable of producing penitrem A. The man suffered from tremors for some hours but recovered fully overnight.

The best examples of toxic reactions are seen in domestic animals. Ingestion of straw with an abnormally high concentration of moulds has resulted in tremors and spasms in swine, often with fatal results. Another acute effect from ingestion of contaminated crops is bleeding from the gastrointestinal tract.

The long-term effects of consumption of low level-infected crops may possibly create a more serious veterinary problem with loss of appetite, depression of growth in nonadult animals, damage to the liver and subsequent death.

Acute reactions to ingestion of aflatoxin in humans were tragically demonstrated in India in 1974, where a severe drought was followed by heavy, out of the season, rainfall prior to the corn harvest. The harvested corn was infected with *A. flavus*, and more than 100

Fig. 45. *Aspergillus flavus* seen through the stereomicroscope, which under special conditions is able to produce the mycotoxin aflatoxin B_1, known to be the most potent carcinogenic substance of biological origin.

Fig. 46. Mouldy Brazil nut infected by different *Aspergillus* species. The risk of ingestion of mycotoxins from mouldy nuts has, in some countries, resulted in import of Brazil nuts without shells for better control of contamination.

people died after eating the mouldy corn. In 1981, 12 people in Kenya died under similar circumstances.

Another important and frequently occurring toxic substance, the carcinogenic ochratoxin A, is produced by *A. ochraceus* and *Penicillium verrucosum*. *P. verrucosum* grows on stored crops and produces the toxic metabolites that make the crop toxic especially to swine and poultry. The manifestations are damage to the kidneys and retarded growth rate. A similar effect on humans is known as the Balkan endemic nephropathy, a condition under current research and discussion. *P. polonicum* and *P. aurantiogriseum*, which are common in the Balkan area and produce some nephrotoxic glycopeptides, may be involved in this disease.

Other toxigenic moulds are species of *Fusarium*, and *Stachybotrys*, which produce macrocyclic trichothecenes with potent effects on the immune system and protein synthesis. *Stachybotrys chartarum* produces at least five macrocyclic trichothecenes which are dermotoxic and cytotoxic.

The *Fusarium* trichothecenes T-2 and deoxynivalenol (DON) are produced in grains, especially those overwintering in the fields in cold climates. The resulting toxicosis after eating the grains is named the haemorrhagic syndrome. The clinical symptoms are nausea, vomiting, necrotic oral lesions, dermatitis and bloody diarrhoea.

During the last years of World War II, hundreds of thousands of people in Siberia experienced this because there were no labourers to harvest the fields, with a subsequent overwintering of the crops. The grains were harvested and eaten the following spring, with fatal consequences.

For many years it was known that *Fusarium moniliforme* (= *F. verticillioides*), an important spoiler of maize, caused severe diseases in horses (encephalomalacia) and possibly oesophageal cancer in humans. A series of toxic or mutagenic compounds, moniliformin and fusarins, were isolated by extraction with organic solvents from this fungus. These components, however, did not appear to cause the above-mentioned diseases. Around twenty years later it was demonstrated that the offending substances were lost in the discarded water fraction which was not analysed, because mycotoxins are usually extracted with organic solvents. The fumonisins turned out to be highly carcinogenic, as demonstrated by animal experiments. Furthermore, fumonisins are potent inhibitors of ceramide synthetase and therefore affect production of sphingolipids in cell membranes.

Fig. 47. DRYES is recommended as a standard medium for the detection of *Penicillium verrucosum*, the only important producer of ochratoxin in wheat and barley. Note the characteristic red brown colour of the reverse side of the Petri-dish.

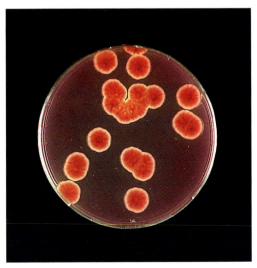

Chemical characterisation

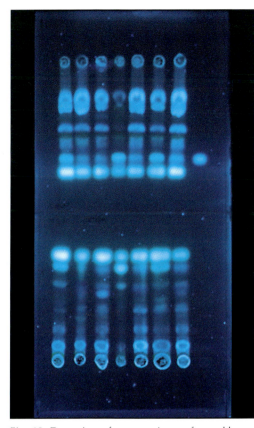

Fig. 48. Detection of mycotoxins performed by means of thin-layer chromatography (TLC), a method where metabolites produced by the fungus are extracted directly from the culture on the Petri-dish and applied to the TLC plate. Identification of the mycotoxins is based on the localisation and colour of the individual spots compared with known standards.

The discovery of aflatoxin B_1 – the most carcinogenic biological substance known today – initiated worldwide interest in isolation and characterization of mycotoxins. Production of mycotoxins is dependent on water availability as well as cold or heat stress. Today approximately 3000 specific metabolites are characterized, which are produced by about 600 different mould species.

The mycotoxins (with various chemical structures) are relatively small molecules (<1000 daltons) compared with, for example, inhalation allergens. The structural diversities explain the different biological effects and the different chemical properties.

Mycotoxins in the indoor climate

Recent research has been concentrated on the respiratory route and direct skin contact with the toxic metabolites. Mycotoxins located inside as well as on the surface of the spores have been demonstrated. This implies that there can be a direct contact to mycotoxins not necessarily prior to ingestion of mouldy foodstuff, in which the mould has excreted the specific metabolites.

Furthermore, it has been demonstrated that the potency of the respiratory route is higher than the alimentary. This means that the dose of mycotoxin required to cause particular effects is typically one order of magnitude less when administered by the respiratory route than by the alimentary route.

Together with the problems caused by ingestion of mouldy foods, emphasis is now very much on health implications of inhalation of mycotoxins from moulds in damp houses. Typical bioindicators for moisture in a building are growth of *Aspergillus versicolor*, *Penicillium chrysogenum* and *P. expansum*. On wet gypsum boards, *Stachybotrys chartarum* is frequently recorded.

Experimental growth on wallpaper-glue agar demonstrated that *A. versicolor* produces the carcinogenic toxin sterigmatocystin, *P. expansum* secretes the nephrotoxic components citrinin and patulin, which affect killer cell (white blood cells active in immune defence system) activity and phagocytosis. Finally, it has been demonstrated that *Stachybotrys* produces the above-mentioned macrocyclic trichothecenes, the most potent inhibitors of protein synthesis, with immunosuppressive effects and with dermotoxic effects. Breathing 1000 *Stachybotrys* conidia/m³ air – regardless of whether alive or dead – may produce different undesirable effects on the immune system and other systemic effects.

I: Aflatoxin B₁

II: Ochratoxin A

III: Fumonisin B

IV: T-2 toxin

V: Communesin A

Fig. 49. Chemical structures of different mycotoxins.
I: Aflatoxin B₁. Important producers are *Aspergillus flavus*, *A. parasiticus* and *A. nomius*. Found mainly on nuts, corn and peanuts.
II: Ochratoxin A, a carcinogenic kidney toxin. Producers are *Aspergillus ochraceus* and *Penicillium verrucosum*. Only the latter species is known to cause mycotoxicosis. Isolated mainly from wheat and barley.
III: Fumonisin B, a carcinogen, which is also toxic to the brain. Produced by *Fusarium* species related to *F. moniliforme* growing mainly on corn.
IV: T-2 toxin is toxic to the skin (dermatoxic) and also affects mucous membranes. Known to be highly immunosuppressive. The toxin is produced by *Fusarium culmorum*, *F. poae* and related species. Found mainly on cereals.
V: Communesin A, a cytotoxic metabolite of still unknown significance, produced by *Penicillium expansum* growing on pomaceous fruits, nuts and on water-damaged building material.

50

Volatile mycotoxins

Some of the specific metabolites from the moulds are released as volatiles into the indoor air. These are called microbial volatile organic compounds (mVOC).

The compounds are organic solvents consisting mainly of alcohols, ketones, hydrocarbons (mono- and sequiterpenes) and aromatics, often with an unpleasant odour, such as the heavily musty smelling 2-methyl-isoborneol from *Penicillium commune*.

Other volatiles characterized are 2-methyl-1-propanol, 3-methyl-1-butanol and 3-octanone isolated from commonly occurring indoor mould species like *Aspergillus* and *Penicillium*.

These observations add a new concept to the handling and investigation of building problems resulting from water damage or other moisture problems.

Ongoing laboratory experiments with volatile and nonvolatile specific metabolites will produce documentation on their damaging effects.

Often the effects are only incompletely studied and understood. With few exceptions, as mentioned in the case studies of acute effects on humans, valid information on acute or long term effects on humans is still lacking.

Elucidation of the toxic effects and risks connected with short- or long-term exposure to toxic mould spores and their volatiles will involve collaboration between researchers from many different disciplines.

Fig. 50. Mouldy corn silage heavily infected with a special kind of *Penicillium roqueforti*. In some cases, mycotoxicosis in domestic animals following intake of such silage has been demonstrated.

Chapter 5

Allergy and other adverse health reactions to moulds

Most filamentous fungi have the ability to produce conidia in large quantities.

When such a massive number of spores are liberated from a growth focus to the ambient air, it can be regarded as organic dust. This dust can, like other types of dust, sediment on surfaces or it could be inhaled by humans and deposited on the mucosal surface of the upper airways and in the eyes. Repeated exposure to large amounts of fungal propagules risks the development of specific allergic reactions.

Two different types of allergy may be the result of repeated heavy exposure to spore dust: Type 1 and Type 3.

Type 1 allergy

In Type 1 allergy, the over-reaction of the immune system will occur after massive and long exposure to the offending agent, ranging from months to years.

However, once the immune system has been triggered, the allergic reaction will be elicited upon exposure to minute amounts of the specific allergen.

Thus, the yearly outdoor-dose of the major allergen (Alt a 1) from the mould *Alternaria* is estimated to be as little as 1 μg.

The frequency of Type 1 allergy to moulds is not as high as to other important inhalation allergens. There are an estimated 8% adult allergic patients and 20-25% children in the population with established airway allergy.

Epidemiological studies point towards mould allergy in children as a transient phenomenon, maybe due to the immature state of the child's immune system.

Development of Type 1 allergy in most cases implies a genetic predisposition. Patients with this disposition have the ability to produce certain antibodies in larger amounts than "normal" people.

The antibodies named IgE-antibodies are produced as a result of a reaction in the body's immune system to protect the individual against intrusion by "foreign substances", the so-called antigens.

Fig. 51. The allergenic proteins from the common mould *Alternaria alternata* demonstrated by means of a technique where serum from an *Alternaria*-allergic patient is used. The patient's specific IgE will bind to the fungal proteins to which he or she reacts. The technique is called crossed radioimmunoelectrophoresis (CRIE).

53

Table 4.
Common airborne moulds

Indoors	Outdoors
Alternaria alternata	*Alternaria alternata*
Aspergillus fumigatus	*Aureobasidium pullulans*
A. niger	*Botrytis cinerea*
A. versicolor	*A. versicolor*
Aureobasidium pullulans	*Cladosporium* spp.
Botrytis cinerea	*Drechslera* spp.
Cladosporium cladosporioides	*Epicoccum nigrum*
C. herbarum	*Fusarium* spp.
Penicillium chrysogenum	*Penicillium chrysogenum*
P. polonicum	*P. polonicum*
Stemphylium botryosum	*Penicillium* spp.
Trichoderma viride	*Trichoderma viride*
Ulocladium spp.	*Ulocladium* spp.

Table 5.
The most frequent moulds found in house dust in samples of homes in Canada (*n* = ca. 50) and Denmark (*n* = 100) (as a percentage)

Canada		Denmark	
Genus	%	Genus	%
Penicillium spp.	80	*Mucor* spp.	98
Rhizopus spp.	73	*Penicillium* spp.	97
Cladosporium cladosporioides	67	*Alternaria alternata*	80
Alternaria alternata	57	*Aspergillus fumigatus**	67
Aspergillus niger	53	*Ulocladium* spp.	67
Penicillium viridicatum	39	*Cladosporium* spp.	62
Mucor spp.	31	*Aspergillus* spp.	50
Trichoderma viride	25	*Rhizopus* spp.	37
Ulocladium botrytis	22	*Trichoderma viride*	20

*After incubation at 37°C.
House dust is a very stable biotope throughout the world. The fungal flora is more or less the same, but quantitative differences have been demonstrated.

The causative agents in respiratory allergy are usually proteins. Where fungal allergy is described, it is caused by proteins from the fungal spores. As protein is water soluble, it is extracted by means of mucosal fluid in the person's upper airways, where the spores have been deposited together with the air inhaled.

Extraction of the protein molecules – the antigens – takes around 30 seconds. After deposition and extraction, the antigenic molecules will penetrate the mucosal barrier and meet "the garbagemen" of our immune system: the macrophages, white blood cells, which engulf and degrade all the foreign substances in our body, including our own dead cells.

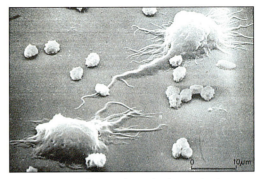

Fig. 53. Macrophages, from the lung, chasing inhaled spores of *Aspergillus fumigatus* in order to engulf and destroy them.

Fig. 52. Moist window-sill with a monoculture of *Cladosporium herbarum*. This window was sited above the bed of a 10-year-old girl with severe asthma (Type 1 allergy) and atopic dermatitis.

During degradation of the allergen, the macrophage also acts as "antigen-presenting cell" (APC), by binding fragments of the fungal spore proteins (peptides) to a certain component in the wall. From this position, the complex will be presented to the other cells of the immune system and, in this way, signal to them that there are foreign components present in the body.

The macrophage may show the fragment of the allergen to another white blood cell: the T-cell. The fragment of the allergen together with the component from the macrophage can be bound to a specific complex present on the surface of the T-cell. This complex is named a T-cell receptor. Many different T-cells exist in the body and react with all possible allergen fragments.

Before the allergen was degraded, another white blood cell, the B-cell, at the same time had recognized the allergen by means of an antibody. Like the T-cell-receptors, a lot of different B-cell antibodies recognizing different antigens are present.

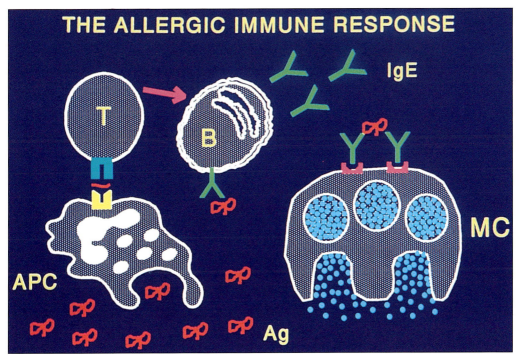

THE ALLERGIC IMMUNE RESPONSE

IgE

MC

APC

Ag

Fig. 54. A schematic outline of the allergic immune reaction. After penetration of the nasal mucosal barrier, the allergenic molecule (Ag) is met by amoeba-like white blood cells (APC = the antigen-presenting cell), the macrophages, which engulf and digest the molecule to form minor fragments. After that, the fragments are bound to receptors in the membrane of the macrophage. The fragment is then presented to T-cells (T), which produce different mediator substances, of which the interleukins are very important. They are able to bring a message to the B-cells (B) to produce IgE antibodies against the specific allergens presented by the macrophages. The specific IgE produced will bind to receptors on the mast cells (MC).

By means of some mediator substances, the so-called interleukins, the T-cells will communicate with the B-cells and tell them to produce IgE antibodies, which act chemically on those specific fungal antigens. The B-cells can be regarded as small and very efficient "antibody factories" in the human body.

In nonallergic people, the above-mentioned reaction between antigens and antibodies takes place naturally, in order to protect the individual against many different kinds of infection, e.g. parasites or bacterial and viral infections.

In allergic people, however, the over-stimulated production of IgE antibody will create an overreaction (hyperreaction) which will induce several of the unwanted effects described later on.

In the case of Type 1 allergy, allergic reactions such as hay fever or asthma will occur within a few minutes after exposure.

Some of the mechanisms behind the

allergic reactions can be explained in the following way.

A very important substance produced during the allergic reaction is histamine, a potent mediator-substance which influences blood pressure, smooth muscles and secretion from the mucosal glands.

Two other types of white blood cell, the mast cell and the basophil cell, are involved in the secretion of histamine. The IgE antibodies produced by the B-cells will attach themselves to the surface of the mast cells (present in the skin) and the basophils (present in the blood). Both types of cell contain many histamine grains, and a reaction between the antigens (e.g. fungal protein) and the IgE antibodies on the surface of the cells will initiate a degranulation of the cells with subsequent liberation of histamine, producing the following effects:

1) Contraction of smooth muscles around the bronchioles (the branches of the bronchial tree) resulting in asthma.

2) Secretion of mucus from the glands in the lungs, resulting in infection and inflammatory reactions of the lung tissue of the asthmatic patient.

3) Secretion of nasal fluids and tears from the eyes, resulting in congestion of the nose and itching eyes.

4) Finally, histamine liberation will influence the vascular permeability of the blood vessels, resulting in an increased permeability of the vessels with the risk of decline in blood pressure. This condition, which very seldom occurs, may induce the so-called "allergic shock" or anaphylactic shock, which may be fatal.

Fig. 55. Mast cells with conspicuous histamine granules. Repeated exposure to the same allergens will make them bind to the corresponding IgE antibodies placed on the mast cells. Hay fever or asthma will follow after liberation of histamine, which has a negative effect on blood vessels, mucous glands and smooth muscles.

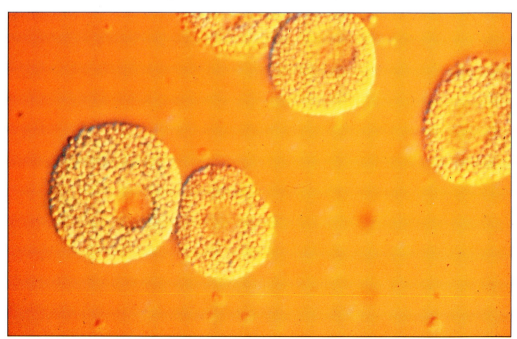

Type 3 Allergy

The other specific allergic reaction to inhalation of moulds or mouldy material is extrinsic allergic alveolitis.

In this type of antigen/antibody reaction, IgE is not involved. The reaction is mediated by other antibodies, mainly IgG, which form immune complexes with the inhaled antigens. These complexes will initiate different inflammatory responses and sometimes trigger the complement system, which may result in asthma.

The condition is generally associated with an occupational disease, which can, for example, affect farmers handling mouldy hay or other crops (farmer's lung), workers occupied with whisky production handling mouldy barley (malt worker's disease), workers removing fungal growth from cheese (cheese washer's lung) or people handling mouldy wood chips for fuel or moist timber (wood trimmer's disease).

The symptoms, which occur 6-8 hours after exposure, are general malaise, flu-like symptoms, elevated temperature, muscle and joint pains, dyspnoea, weight loss and later asthma, leading to fibrosis of the lung tissue.

Cessation of exposure to the offending agents, before the onset of fibrosis of the lung tissue, will normally lead to a return to healthy state.

Fig. 56. Handling mouldy hay may cause a delayed allergic reaction 4 to 6 hours after exposure. This condition was described by the Italian medical doctor Ramazzini in 1713 and is recognised as the first occupational disease. Inhalation of organic dust from poorly ventilated hay or grain contaminated by growth of, for example, thermophilic actinomycetes and species of *Aspergillus* may cause general malaise, flu-like symptoms such as fever and muscle pains, dyspnoea and later on asthma. This is called farmer's lung disease or thresher's lung. The condition described could also be attributed to the so-called organic dust toxic syndrome (ODTS), which is considered to be toxic rather than allergic as no specific antibodies can be demonstrated.

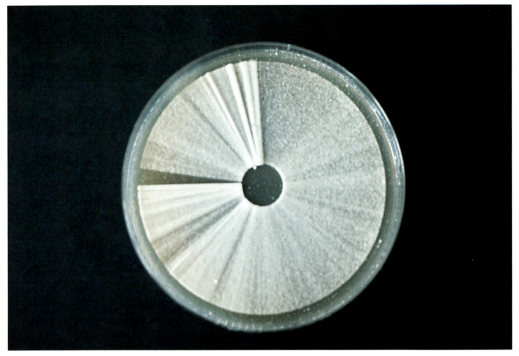

Fig. 57. Culture plate on which active sampling of airborne conidia from the cheeses has taken place. Using the BIAP slit sampler method, the highest possible number, approx. 6000/ m³ of air, has been recorded.

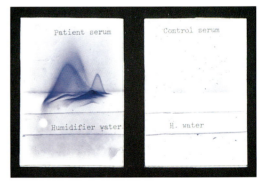

Fig. 58. Antibodies from a patient with humidifier disease working in a printing office against infected water taken from the humidifier. To the right, a control person not exposed to the infected water, who will not develop such antibodies.

Fig. 59. Worker in a distillery maltings handling mouldy malt infected with *Aspergillus clavatus.* Malt worker's lung, a classical manifestation of extrinsic allergic alveolitis, may develop.

60

Fig. 60. Brie cheeses in a storage room. After 6 to 7 days *Penicillium camemberti*, which covers the surface of the cheeses, will liberate conidia in huge quantities to the ambient air, with a subsequent risk of respiratory problems among the cheese workers.

Fig. 61. Humidifier water may be a reservoir for many different biological organisms like those seen on this culture plate. Gram-negative bacteria will be the first colonisers; then yeast, other bacteria, moulds and sometimes amoebae will follow. Inhalation of aerosols containing cells, fragments of organisms or their metabolic products may cause humidifier disease, an allergic/toxic condition.

Table 6.
Important moulds causing occupational respiratory allergy

Allergen source	Environmental exposure	Disease
Alternaria alternata	mouldy logs	wood pulp-worker's lung
Aspergillus clavatus	mouldy malt	malt worker's lung
A. fumigatus	mouldy wood chips mouldy sphagnum	wood chip burner's lung greenhouse lung
A. niger	enzyme	asthma
A. oryzae	acid production	asthma
Botrytis cinerea	mouldy plants	asthma, greenhouse lung
Eurotium rubrum (*Aspergillus umbrosus*)	mouldy hay	farmer's lung
Penicillium camemberti	cheese production	asthma, cheese worker's lung
P. commune	mouldy cheese	cheese washer's lung
P. roqueforti	cheese production	asthma, cheese worker's lung
Rhizopus stolonifer	sawmills	wood trimmer's disease
Scopulariopsis brevicaulis	mouldy tobacco	tobacco worker's lung
Mixed mould populations	ventilation ducts mouldy garbage and compost for recycling humidifiers	waste-handling disease* humidifier disease

The specific names of the diseases all cover the same diagnosis: Type III allergy. This is also called extrinsic allergic alveolitis (American: hypersensitivity pneumonitis)
* ODTS (organic dust toxic syndrome)

Sick-building syndrome

Health complaints from people occupying nonindustrial buildings with heavy personal traffic such as schools, kindergartens or public offices have been increasing during the last decade.

As an astonishingly similar pattern of symptoms ranging from mucosal complaints to general symptoms such as headache and extreme fatigue are displayed by people working in such buildings in the Western hemisphere, WHO decided to list this collection of symptoms and call them the sick-building syndrome. The reasons for these symptoms are multifactorial.

Elevated temperatures caused by too many personal computers installed in rooms which were either too small or inadequately ventilated have been one of the reasons for the lack of well-being.

Complaints from the occupants will be forthcoming as a result of emission of volatiles from building material or from materials used in the offices such as sulphur compounds from carpets, formaldehyde from moist chipboards or components from paint.

Accumulation of dust mainly due to inadequate or reduced cleaning is another important reason for indoor climate complaints. However, indoor air quality may also be influenced by pollution from micro-organisms.

Damp-building syndrome

Micro-organisms, including moulds, will always be present indoors, partly from outdoor sources and partly liberated from the surface of the skin or clothes and shoes of people present in the rooms. This is the normal flora.

In buildings with moisture problems or in water-damaged buildings, the number of micro-organisms can be considerably higher and the composition of the microbial flora often quite different.

Generally being present in humid or water-damaged buildings very soon creates problems for allergic patients, who could be called "human smoke detectors". Their hyperreactive airways make them extra susceptible to emission of both odours and particles, with subsequent early and pronounced reactions to the actual irritants compared with a nonallergic person.

A prolonged water-damage problem may, however, also cause troubles for nonallergic people, who may develop several of the mucosal and general symptoms described in the above-mentioned WHO list from 1981 on the sick-building syndrome.

Published case histories, with confirmed diagnosis on building-related adverse health reactions, are very seldom reported, apart from reports on specific allergies.

Reports encountered daily at work and from colleagues show a similar pattern of complaints from people occupying or living in damp houses with mould growth.

The most frequent mucosal and general complaints are listed in Table 7.

Susceptible persons can take a few breaths before the symptoms are present, and they experience an ever decreasing limit to the time that they can comfortably spend in a sick building.

The element of sensitization demonstrated in the above-mentioned decreasing time for stay in a mouldy building could be attributed to the so-called "specific chemical hyper-sensitivity", which may include immunological reactivity and toxic sensitivity. Further

Table 7.
Frequent complaints in moist rooms

Mucosal symptoms	General symptoms
Itching eyes	Headache
Difficulties with contact lenses	Extreme fatigue
Blocked nose	Lack of concentration
Sore throat	Lack of memory
Hoarseness	General malaise
Burning sensation of the skin	Lethargy
Recurrent sinusitis	Dizziness

epidemiological studies of, e.g. water-damaged kindergartens have demonstrated statistically significantly higher infection rates, especially in connection with upper airway infections.

The possible mechanisms for development of these symptoms mentioned above seem to be multifactorial. Recently hypotheses perhaps also point towards irritative, toxic and immunological influences from different types of component such as fungal allergens, mycotoxins, endotoxins (components from gram-negative bacteria) and microbial volatiles.

In fact, there are different adverse health reactions to a large number of biologically very active components.

As long as the evidence for the connection between environmental findings and clinical symptoms is not sufficiently definite, working hypotheses should include the possible different immunological and toxic effects.

Future research must be aimed at an elucidation of the pathophysiological mechanisms evoked by exposure to micro-organisms and their emissions, and the subsequent changes in people.

Chapter 6
Fungal infections

Fungi can act as causative agents for adverse health effects in different ways. They can be the cause of airway allergy, they can cause irritation of eyes, nose and throat and they can have toxic effects on many different organs after intake of mouldy foodstuffs or after inhalation of mould spores. Some moulds may cause skin irritation after direct contact with the skin.

Like the bacteria and the viruses, fungi can cause infections, i.e. be pathogenic.

Systemic fungal infections are, however, much more rare than bacterial and viral infections.

Very few fungal species cause systemic infections in human beings, the so-called mycoses.

Mycoses may be divided into opportunistic and nonopportunistic infections. The opportunistic ones occur in patients with lowered resistance to infections, e.g. patients with a compromised immune system such as diabetic patients

Fig. 62. *Candida albicans* usually forms separate yeast cells, but pseudomycelium may occur.

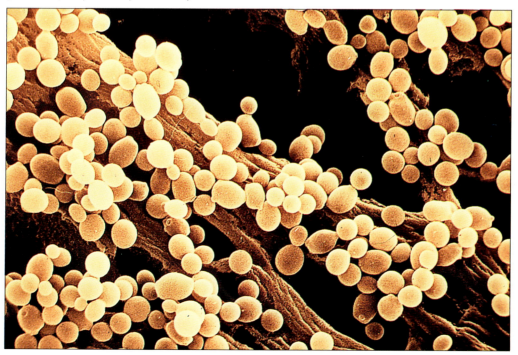

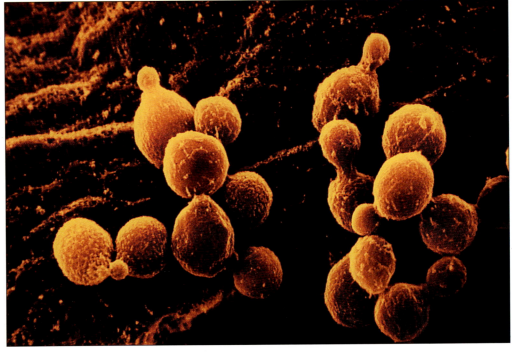

Fig. 63. SEM of *Cryptococcus neoformans,* a yeast causing lung infection.

or AIDS patients or patients undertaking chemotherapy. The yeast *Candida albicans* is an example of such an opportunistic fungus.

Nonopportunistic infections may be caused by *Cryptococcus neoformans, Histoplasma capsulatum* and *Coccidioides immitis.*

Recently it has been found out that *Cryptococcus neoformans* can be isolated as two different serotypes: type AD, isolated from pigeon droppings, and serotype B, found in connection with two different species of *Eucalyptus,* both of Australian origin.

Histoplasma and *Coccidioides* are most often isolated from very dry regions in North America, especially in Arizona and New Mexico. They all cause pulmonary infections which can be misdiagnosed and classified as other types of lung disease.

Another group of pathogenic orga-

Fig. 64. SEM of a typical macroconidium of *Histoplasma capsulatum.* The fungus is isolated from the dung of birds and bats. It is especially common in parts of North America, but is also found in South America and the Far East. *H. capsulatum* may also cause pulmonary infection following inhalation of airborne microconidia from the fungus.

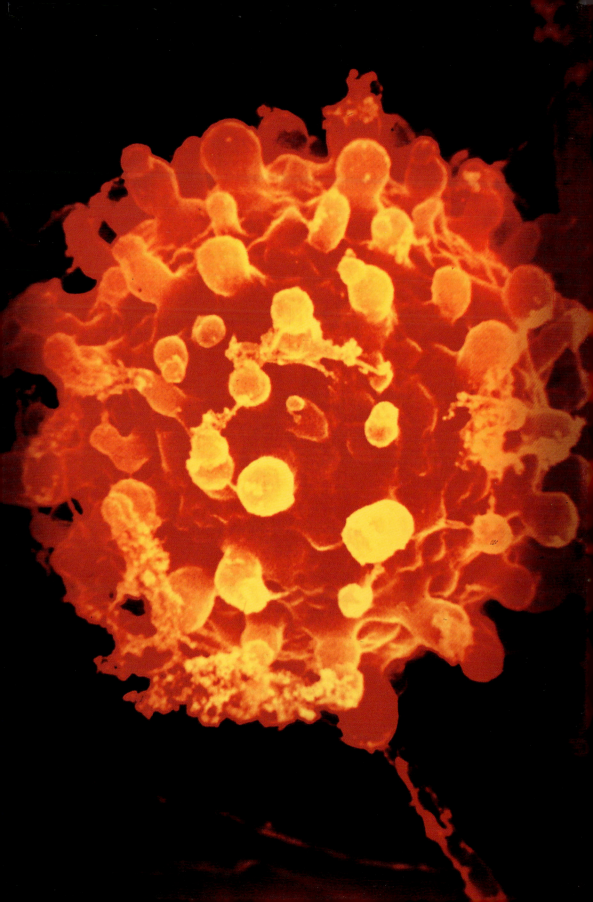

Fig. 65. Calf with *Trichophyton verrucosum*. The fungal infection is easily transferred to the people taking care of the animals.

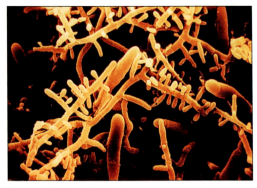

Fig. 67. SEM of *Trichophyton mentagrophytes* with macro- and microconidia. The fungus is often isolated from infections between the toes.

nisms is the dermatophytes (derma = skin), which attack skin, hair and nails; species of the genus *Trichophyton* are frequently involved.

T. rubrum and *T. mentagrophytes* attack the nails and especially the skin between the toes, but seldom hair.

Although there is no established evidence, it might be that the increasing use of rubber footwear as well as extensive use of jogging footwear could result in a more tight and closed envi-

Fig. 66. Culture of the dermatophyte *Trichophyton mentagrophytes* grown on Sabouraud dextrose agar, a standard medium for dermatophytes.

Fig. 68. SEM of *Microsporum gypseum* with macroconidia.

ronment for the feet, which could lead to a higher frequency of mycoses connected with the feet. Furthermore, infections with dermatophytes are often picked up in public swimming pools or sport facilities.

Another dermatophyte, *Microsporum canis*, may be contagious following contact between humans and infected cats, dogs and calves.

This fungus attacks the hair but seldom nails or skin.

The ubiquitous mould *Aspergillus fumigatus* may also act as a pathogenic organism infecting particularly the lungs in the immunocompromised patient. In patients undergoing long and intensive antibacterial medication, e.g children with cystic fibrosis, *A. fumigatus* can be isolated from sputum (saliva) and the airways. Lung tissue may also be invaded by this fungus, where a so-called aspergilloma (a fungus ball) may be the result.

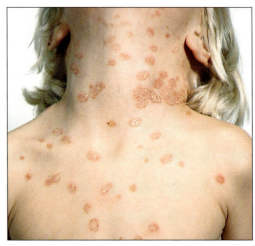

Fig. 69. Girl with ringworm, a skin infection caused by the dermatophyte *Microsporum canis*. Such infections are often transferred to human beings from cats or dogs.

Fig. 70. A rare infection of the skin with the ubiquitous filamentous mould *Aspergillus fumigatus*, one of the few pathogenic filamentous fungi. This infection, a so-called aspergilloma or fungus ball, is known to occur in a pre-existing cavity in the lungs, e.g. after tuberculosis and almost always in connection with deficiency of the immune system.

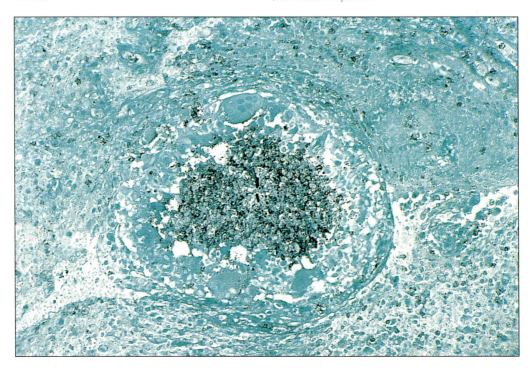

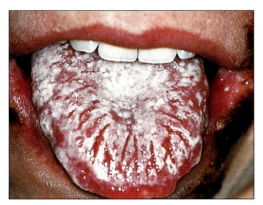

Fig. 71. Oral candidosis – thrush, a common manifestation of *Candida albicans* infection in a patient with a compromised immune system, in this case an AIDS patient.

trolled by drugs until the patient's immune status improves. Then it is possible to tackle, and hopefully cure, the fungal infection.

For some infections, long treatment is needed and some of the drugs used have rather unpleasant side effects.

Effective medication for fungal infections is of increasing importance. One reason for this is the extended use of organ transplantation, in which immunosuppressive treatment of the patient is necessary to avoid rejection of the transplant.

Suppression of the immune system increases the risk of infections, including fungal infections, and AIDS patients, who are also a high-risk group for fungal infections, require effective medication.

Medical treatment of systemic mycoses is often more complicated than in the case of bacterial infections. This is because of an interaction between fungus and host. The immunocompromised status of the patient makes it is difficult to obtain an effective treatment.

Whereas bacterial infections are dealt with by the part of the body's immune system that produces the antibodies (the humoral immune response), the cellular immune response deals with the fungal infections. If either the macrophages or the T-cells are malfunctioning, which can be one of the reasons for fungal infections, the infection must be con-

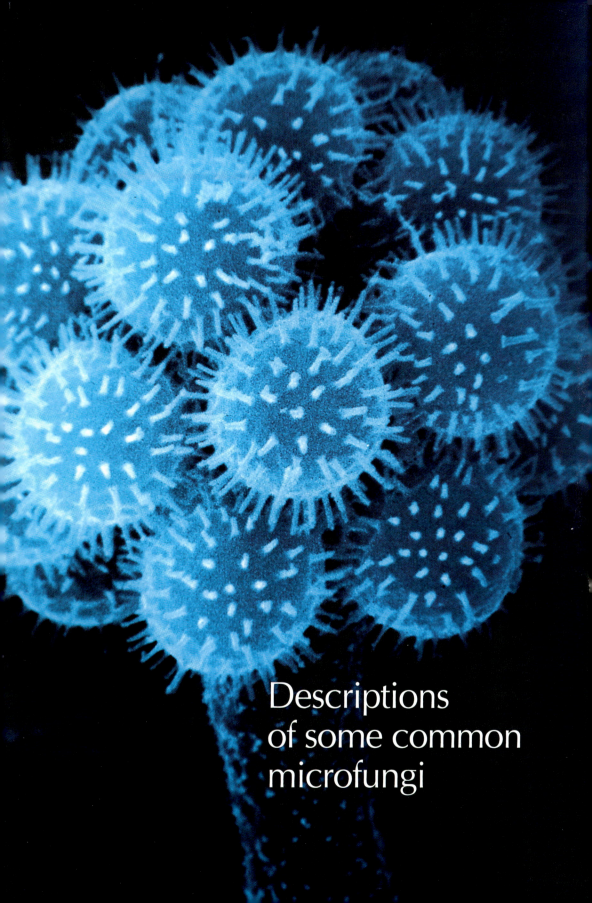

Descriptions
of some common
microfungi

Alternaria alternata

From Latin *alternare*: alternate,
or from Greek *alteres*: a sort of dumb-bell, a short bar
of wood or iron used for exercising the muscles.
Synonym: *Alternaria tenuis*.

Description

When grown on malt extract agar (MEA) at 25°C, the colonies reach a diameter of 6 cm in 7 days. Colonies are black to olivaceous-black or greyish.

Conidia are very characteristic. Shape ovoid to pear-shaped or ellipsoidal with a beak not exceeding one third of the conidial length. 18-83 × 7-18 μm in size. Colour pale to medium brown, either smooth-walled or verrucose with several transverse and longitudinal septae.

Difficulties with maintenance of a culture of this species in the laboratory are often experienced, as several transfers of a strain will lead to a decrease in sporulation, resulting in sterility. However, cultivation on media which contain only a few nutrients often helps to develop its typical structures.

Ecology

A. alternata is an extremely common saprophyte found on plants, foodstuffs and textiles and in different types of soil. Decaying wood, wood pulp and compost are likewise favourite substrates for this species.

Temperature requirements for growth are minimum 2°C, maximum 32°C, and

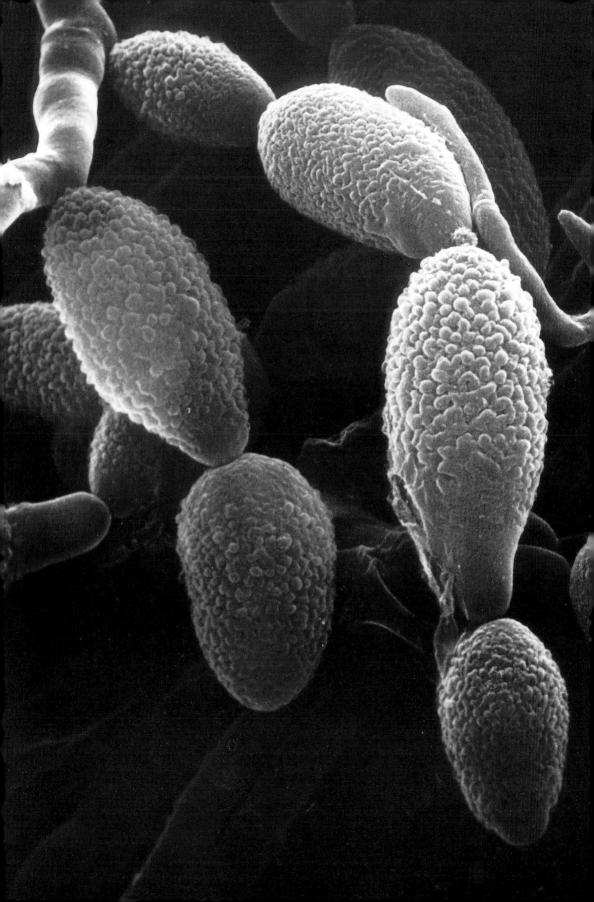

the optimum temperature lies between 25 and 28°C.

This common mould has a worldwide distribution, and the conidia are frequently recorded from outdoor air, where they reach their peak during late summer – often in high concentrations. Indoors it is found in air, in house dust, on damp spots such as window frames with condensation and on humid walls or ceilings.

In greenhouses where flowers such as *Chrysanthemum* or tomatoes are grown, *A. alternata* will often be found on sick or dead plants because of its tendency to inhabit substrates such as rotten organic material.

Macroscopically, the colonies of other common dark-spored moulds as *Stemphylium, Ulocladium, Helminthosporium* and *Curvularia* look very much like those of *Alternaria*. Their morphology is, however, quite different.

Practical application

A. alternata is cultivated for production of biomass, used as source material for allergen extracts for diagnosis and treatment of airway allergy to *Alternaria*.

Damaging effects

Agricultural aspects: *Alternaria* is able to damage cereals and vegetables under field conditions but rarely during storage. Black spots on tomatoes, onions and carrots may often be caused by *Alternaria*. Tobacco and cotton leaves may also be attacked. Some very potent enzymes, especially cellulases, easily decompose the cellulose components of the material invaded.

Medical aspects: The relatively large abundance of *Alternaria* conidia in outdoor air and its occurrence in mouldy houses makes this mould one of the most important fungal allergen sources. The major allergens have been isolated and characterized, and it has been shown that *A. alternata* conidia wherever they have been isolated in the world always contain the same allergens. Furthermore, patients with airway allergy to *Alternaria* react against the same proteins (allergens) regardless of the geographical origin of these conidia.

Closely related genera such as *Ulocladium* and *Stemphylium* contain allergens identical to those found in *A. alternata*.

Allergy to *Alternaria* will often cause asthmatic reactions of the immediate type (IgE mediated). Baker's asthma is considered to be connected with inhalation of *Alternaria* conidia present in flour. Two cases of the farmer's lung type of allergy (IgG mediated) due to *Alternaria* were recently reported in two children living on a farm.

Infection with *Alternaria* is seldom reported. *A. alternata* can cause skin disease, e.g. of the toe. Infection of the brain has been described, but these cases are extremely rare.

Specific metabolites

Mycotoxins from *Alternaria* have not yet been subject to much investigation. However, *A. alternata* produces alternariol, which also has antifungal effects. Other specific metabolites are alternariol monomethylether (AME), altertoxin I and II (mutagenic toxins), altenuene, altenusin and tenuazonic acid. These mycotoxins may be produced in tomatoes, apples, olives, wheat, *Sorghum,* sunflower seeds and pecans. A feed sample involved in mycotoxicosis in Australian broiler chickens in fact contained 10 mg/kg alternariol and 3.6 mg/kg AME. It has been suggested that *A. alternata* is associated with oesophageal cancer in Linxian, China.

Tenuazonic acid is an important mycotoxin that is also produced by another fungus, *Phoma sorghina.* This acute toxin has been implicated in onyalai, a blood disease among people in South Africa.

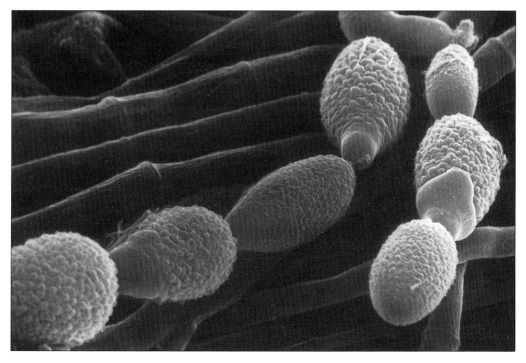

Aspergillus flavus

From Latin *aspergillum*: holy water stoup, *flavus*: yellow.

Description

Relatively fast growing colonies on Czapek agar at 25°C consist of a dense felt of yellow-green conidiophores. Colonies become dark yellow green with age. Faster growth occurs on MEA.

Conidia are globose to subglobose, 3.5-4.5 μm in diameter, pale green, finely roughened. Sclerotia are often produced in fresh isolates – first white then becoming brown to black.

This species has quite a similar morphology to *A. oryzae*, a fungus which is widely used for biotechnology.

Ecology

A. flavus has a worldwide distribution but is isolated mainly from tropical and subtropical regions. It is found on a variety of substrates from foods to wood exposed to sea water, wood pulp, in birds' nests, on leather and cotton and on building materials.

The temperature which permits growth ranges from 17 to 48°C; the optimal growth temperature has a broad range: 25-42°C.

Practical application

A. flavus, especially isolates resembling its domesticated form *A. oryzae*, are claimed to be used for production of

different enzymes such as amylase, keratinase, lipase, protease, fibrinolytic enzymes and urate oxidase.

It is able to degrade carboxymethyl-cellulose and n-alkanes and it produces L-malic and succinic acid.

Damaging effects

Agricultural and medical aspects:

A. flavus has gained a bad reputation due to its ability to produce the aflatoxin B_1 – the most carcinogenic substance of biological origin. The discovery of these fungal toxins is described in Chapter 4.

Aflatoxins are produced worldwide, especially in peanuts, cottonseeds and Brazil nuts. Various factors restrict the production of aflatoxins, e.g. the nature and moisture content of the substrate and temperature.

As described elsewhere, production of the toxic metabolites (particularly afla-toxin B_1 and cyclopiazonic acid) has consequences for the health of domestic animals because both liver and kidneys may be affected after ingestion of moul-dy feedstuff.

Infection: *A. flavus* may cause in-fection of the ear and eye. Infections of lung, heart and bladder have been re-ported rarely. Pulmonary infections in birds have been described occasionally.

Specific metabolites

Substances produced by *A. flavus* other than aflatoxin B_1 and B_2 are cyclopia-zonic acid, kojic acid and aspergillic acid. Examples of volatile compounds produced are 3-methyl-butanol, 3-octa-none, 1-octen-3-ol and ethylene.

The closely related species *A. parasit-icus* has very rough and darker green conidia and produces all known aflatox-ins (B_1, B_2, G_1, G_2) but <u>not</u> cyclopiazonic acid. The species is less common than *A. flavus*.

Aspergillus fumigatus

From Latin *fumigatus*: smoky.

Description

Colonies on CYA have a characteristic dense appearance with a blue-greyish colour intermixed with colourless aerial hyphae.

Growth of colonies will attain a diameter of around 5 cm within a week when incubated at 25°C.

Growth is faster at 37°C. Cultivation on MEA will result in heavier sporulation.

Conidiophores are short and green bearing typical columnar conidial heads. Conidia are globose to subglobose, 2.5-3 μm in diameter, dark green, rough-walled to echinulate.

Ecology

A. fumigatus is a successful saprophyte and has a worldwide distribution. House dust is an ecological niche for *A. fumigatus* and is a habitat from which it has been demonstrated in almost all parts of the world.

The species is thermotolerant and is able to grow in the range between 12 and 57°C. Optimal temperature for growth is between 37°C and 43°C. It is able to grow in an atmosphere of 100% N_2. 10% CO_2 is also tolerated in vitro. It is able to survive pasteurization to 63°C for 25 min and causes heating in hay or cracked corn, resulting in elevated temperatures up to 50°C. Substrates containing different sugars, peptides and amino acids support good growth.

A. *fumigatus* occurs in outdoor and indoor air, different types of soil and on decaying plant material, feathers and birds' droppings, compost, wood chips, self-heated hay and crops.

Organic acids and many different enzymes are produced, including proteases.

Practical application

A. *fumigatus* has been used for production of protein from starchy substrates, including cassava. Single-cell protein and other biomass production by A. *fumigatus* should, however, never be recommended, as this mould is a producer of large amounts of small and easily dispersed conidia, which have pathogenic and allergenic potential. Furthermore, strains of this fungus are capable of producing a large number of potent mycotoxins.

Damaging effects

Agricultural aspects: A. *fumigatus* is able to attack and damage a wide range of stored fruit, crops, nuts and hay. It can spoil the taste of cocoa beans.

In tropical areas it often causes

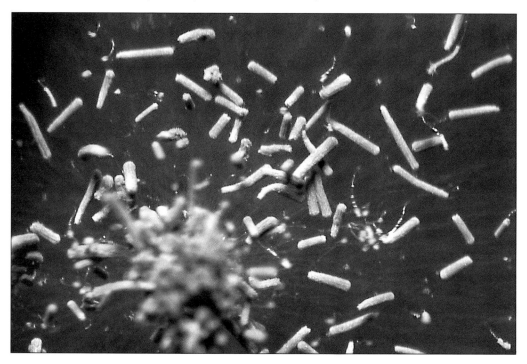

deterioration of cotton and woollen fabrics. Weight loss after deterioration has been noticed in wood from different trees such as birch, spruce and pine. Fuel oil, DDT and PVC materials are degraded.

Medical aspects: A. fumigatus is very important as the infective organism causing systemic mycoses in humans and in domestic animals. In humans, infections frequently occur in the immunocompromised patient, either due to cancer therapy, steroid treatment or diabetes. HIV-positive patients may also be targets for infection with A. fumigatus. Infections are seldom acute, but the cardiovascular and urinary system as well as the brain may be attacked. Occasionally it is found in the respiratory tract of horses. Isolates from expectorates from patients with cystic fibrosis occur occasionally, and it can be found in the human ear and eye.

Asthma and rhinitis (Type I allergy) caused by A. fumigatus are reported. Extrinsic allergic alveolitis (farmer's lung) may develop after exposure, e.g. by handling mouldy hay or wood chips for fuel. The species can cause allergic bronchopulmonary aspergillosis and aspergilloma, two conditions where it acts as a parasite of the lungs – often in the caverna. The first condition may develop a Type I allergy in addition to the growth. The second results in development of a sort of a mycelial ball in the lung tissue. Both illnesses are visible on X-ray radiographs.

Specific metabolites

A. fumigatus produces a large number of important specific metabolites. Compounds with antibiotic activity are gliotoxin, spinulosin, phyllostine, fumigatin, trypacidin, fusigen, ferricrocin and fumigacin (also called helvolic acid, which has antibiotic activity against both gram-positive and -negative bacteria). Other biologically active compounds include fumifungin, a new effective antifungal drug, and fumagillin, active against amoebae.

Fumigaclavine, festuclavine, chanoclavine, sphingofungins, fumitremorgins, verrucologen, tryptoquivalins and fumitoxins are often produced; most of them are toxic, causing death of chickens, tremors (shivering) in various animals, nephrotoxicity or have haemolytic activity (destroy red blood cells).

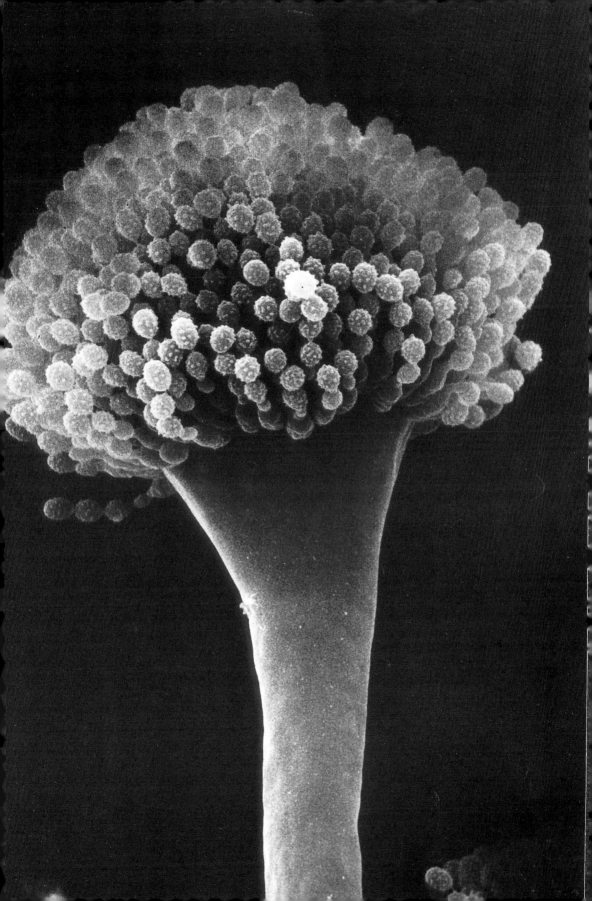

Aspergillus niger

From Latin *niger*: black.
Synonyms: *A. awamori, A. ficuum, A. foetidus, A. phoenicis,*
A. usamii, A. tubingensis, A. pulverulentus, A. nanus, A. intermedius.

Description

The colonies are fast growing on all
substrates. If grown on CZA at 25°C, the
colonies will consist of a white or yellow felt with a compact layer of brown
to black conidiophores and a cream to
yellow reverse side. When grown on
MEA the colonies are thinner but heavily sporulating.

Conidia are globose to subglobose
3.5-5 μm in diameter; brown with
warts, spines or ridges. Strains will vary
in colour from dark brown or purplish
brown to black.

Large cream to buff sclerotia are occasionally produced.

Ecology

A. niger is cosmopolitan and of very
common occurrence. It is often isolated
from house dust, soil, plant litter, dried
nuts, fruits and seeds and different kinds
of untreated textile material such as jute,
hemp and cotton bracts. Hence the
often abundant presence of this species
in the textile industry.

Optimal temperature for growth is
20-40°C. Good growth at 37°C.

Practical application

A. niger is used in the degradation of complex organic materials and waste, such as squeeze remains from production of apple juice, potato garbage and wastewater from processing sugar beets and beetroots and from beer production.

This fungus is an industrially important micro-organism because it is able to decompose plastic and cellulose and is widely used in the production of different enzymes, including amylase and amyloglucosidase, for use in the bread and beer-making industry. It produces organic acids such as oxalic acid and fumaric acid, and the annual fermentation of citric acid is estimated as 5850 million litres worldwide, thus demonstrating the enormous economic significance of this fungal species. A. niger is a GRAS (generally regarded as safe) organism.

Damaging effects

A. niger is widespread and can infect meat and eggs and also cause progressive spoilage, e.g. of cakes and salami. Different spices and sun-dried fruit may contain A. niger. The species is known as the "black" mould of onions and can cause enormous spoilage during transport and storage, especially in warmer climates.

Medical aspects: The fungus can infect the human ear and be the cause of airway allergy. The extensive use of A. niger in the industrial production of compounds such as citric acid and enzymes exposes workers to high concentrations of conidia or proteinaceous fractions of the fungus, with possible sensitisation and different types of airway allergy as an unwanted consequence.

Specific metabolites

A. niger produces a variety of secondary metabolites. Two of them, malformin C and some of the naptho-γ-quinones, have shown toxic effects, but they do not really fall within the definition of mycotoxins given above. Other metabolites are asnipyrone A & B, aspergillin, asperrubrol, asperenones, aurasperones, 4-hydroxymandelic acid, orlandin, nigerazines, nigragillin, orobols and pyrophen. From sclerotia some specific metabolites can be isolated, such as tubingensin A & B and aflavinines.

Aspergillus ochraceus

From Latin *ochraceus*: ochre-coloured.

Description

Colonies on CZA will reach a diameter of 2.5-3.5 cm in one week when grown at 25°C. They consist of a dense ochre-yellow felt with globose conidial heads often splitting into compact columns.

Conidia globose to subglobose 2.5-3.0 μm in diameter are finely rough-walled. Irregular shaped sclerotia – white, later lavender to purple – are often present.

Ecology

A. ochraceus is widely distributed in subtropical and tropical areas but is also found in Europe in different soils. It is isolated from air and house dust sam-

ples and found in fields treated with sewage sludge. It attacks different sorts of grain such as rice, corn and wheat.

Temperature for growth ranges between 12 and 37°C. Optimum temperature is 27°C.

A. ochraceus can produce enzymes for degradation of protein, lipopolysaccharides and pectin. It tolerates salt concentrations up to 30%.

Specific metabolites

Even though ochratoxin A is named after *A. ochraceus*, this important mycotoxin has not been reported to be produced naturally by any *Aspergillus* species in feeds or foods. The laboratory production of this toxin requires certain conditions concerning substrate, age of contamination, pH and temperature. Ochratoxin found in cereals and meat products is produced by *Penicillium verrucosum*.

Other specific metabolites found are penicillic acid, 4-hydroxymellein, flavacol, neo-aspergillic acid, the kidney and liver toxins xanthomegnin and viomellein and the volatile compounds 1-octen-3-ol and 2-octen-1-ol.

Practical application

It is used for biological control of *Fusarium* crown rot of tomato and *Verticillium* wilt and for hydroxylation of sterols and other compounds. Furthermore, it is applied in the transformation and degradation of n-alkanes.

Damaging effects

It spoils tobacco, wood pulp, cotton and leather. Attacks insects such as silkworm and Colorado beetles. Infection route is the hind gut. Isolated from human sputum.

Aspergillus oryzae

From Latin *oryza*: rice.

Description

Colonies on CZA reach a diameter of 4-5 cm within 7 days.

The conidial heads are radiate, pale greenish yellow, later becoming light to dull brown.

The conidia are ellipsoidal at first, later on globose to subglobose, 4.5-8 μm in diameter, but often not uniform in size, smooth to finely rough-walled.

A. oryzae can be confused with *A. flavus*, but typical colonies have a lighter green colour, the conidia are less ornamented and the conidiophore stipes are longer. There are several biochemical differences, for example, it can be differentiated from *A. flavus*, which produces aspergillic acids.

Ecology

A. oryzae has only been found in industrial plants and their immediate surroundings.

Optimal growth temperature is between 32 and 36°C. Growth may be affected by the presence of *Streptomyces* species.

Practical application

A. oryzae has been used for many centuries in Japan and other Asian countries for the fermentation of soybeans to soya sauce and similar products, and is a GRAS organism.

Soya sauce, miso (a paste of soybeans and rice) and saké (rice wine) are produced by means of a starter named koji, which today is produced in modern factories with highly controlled production and selected strains. The ability of *A. oryzae* to produce amylase, amyloglycosidase and many other enzymes is commercially exploited, and these compounds are produced in large quantities. The production of a heat-stable lipase for amending detergents in order to remove fat containing spots from clothes has been successfully achieved. The organism used is a genetically engineered strain of *A. oryzae,* with the gene coding for the lipase originating from the thermophilic fungus *Thermomyces lanuginosus.*

Damaging effects

It has been claimed to occur in brain tissue, but the fungus is not regarded as pathogenic.

Because it is commercially exploited as an enzyme producer, *A. oryzae* can be an occupational inhalation allergen.

Specific metabolites

A. oryzae can produce the following compounds: kojic acid, oryzacidin, the volatiles 1-octen-3-ol, 3-octanone and 3-methylbutanol and an unknown substance with a stale odour. Some isolates may produce the toxins cyclopiazonic acid and/or ß-nitropropionic acid. Strains able to produce the latter two mycotoxins in foods or industrial products are not used.

Aspergillus terreus

From Latin *terra*: earth.

Description

Colonies on CZA will reach a diameter of 3.4-4.0 cm within 7 days at 25°C. On MEA agar it will grow faster and sporulate more densely. The colour of the colonies is yellow-brown, becoming darker with age.

The conidiophores are hyaline and smooth-walled with columnar heads. Conidia are globose to ellipsoidal 1.5-2.5 μm in diameter, smooth and light yellow.

Ecology

This mould occurs in tropical and subtropical zones and has a worldwide distribution in different soils. One very common habitat is the rhizosphere (root zone) of plants. It is recorded from potato, cotton and jute, and it is also isolated from air and house dust.

Growth temperature ranges from 11 to 48°C with an optimum at 37°C.

Practical application

A. terreus produces itaconic acid and mevinolin. Mevinolin, which is able to lower the blood cholesterol levels, was the first compound of its kind approved by the U.S. Food and Drug Administration (FDA). It is sold under the trade names Mevacor® or Lovastatin®.

Damaging effects

Agricultural aspects: *A. terreus* is common on many kinds of seed when stored at high moisture levels. It can cause rot in apples and damage cotton fabrics and leather in the tropics.

   Medical aspects: It is isolated from expectorates from patients with cystic fibrosis, can be found as a parasite in the human ear and is able to attack human skin and nails. It can also be found in lungs of small mammals.

Specific metabolites

A. terreus can produce a large number of specific metabolites, including the nephrotoxin citrinin, the neurotoxin citreoviridin, patulin, terrein, terreic acid, asterric acid, aspterric acid, questin, emodin, sulochrin, erdin, geodin, 6-hydroxymellein, aranotins, aspulvinones, asteroquinones, mevinolins, quadrone, terretonin, terramides, cytochalasin E, astechrome, terrecyclic acids, astepyrone and several other compounds. The only known tremorgenic mycotoxins without nitrogen in the molecule, the territrems, have been found only in one strain of *A. terreus*.

   A. terreus produces substances which are able to inhibit growth of different bacteria and of the dermatophytes *Trichophyton mentagrophytes* and *Sporothrix schenckii*.

   Extracts of corn contaminated with *A. terreus* cause death of mice and of HeLa cells (human cell-lines used in culture).

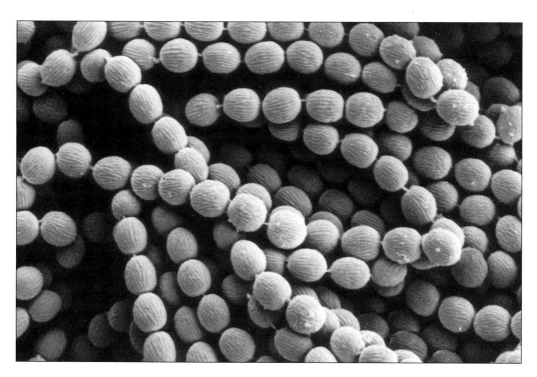

Aspergillus versicolor

From Latin: *versicolor*: multi-coloured
or changing colours.

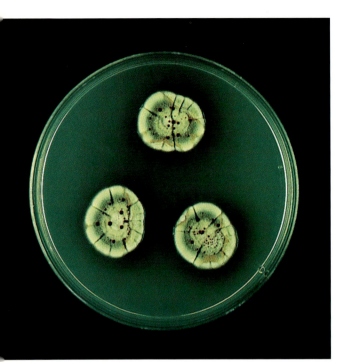

Description

Colonies on CZA are very slow growing
reaching only 1-1.5 cm in 7 days when
grown at 25°C.

The colour of the colonies is at first
white, then more yellow changing to
orange-yellow, to yellow-green often
intermixed with fleshy to pink colours.
On MEA, the colonies will grow faster
and have darker green shades.

The conidia are globose, 2-3.5 μm in
diameter and echinulate, emerald green
in lactophenol.

Ecology

While many *Aspergillus* species are
predominantly found in warm regions,
A. versicolor can also occur in temperate climates. It is found on food products, especially cheese, and in air and
house dust.

Minimum temperature for growth is
4°C, and the optimum temperature is
22-26°C. Maximum temperature for
growth is 40°C.

The fungus is rather xerophilic, and
growth is readily demonstrated on low
water activity media (DG18 agar or MEA
+ 30% NaCl or 40% sucrose).

Practical application

A. versicolor is able to utilise hydrocarbons from fuel oil. It possesses enzymes
for degradation of starch, cellulose,

lipids and proteins. It is able to transform progesterone to testosterone and other closely related ketones.

Damaging effects

Together with other common moulds such as *Penicillium chrysogenum*, *P. expansum*, *P. polonicum* and *Stachybotrys chartarum*, *A. versicolor* is an indicator organism of moisture problems in houses, as it is frequently isolated from water-damaged building materials such as wooden floor constructions, wallpaper and wet mineral fibre boards used for insulation and subjected to long-term humidification.

Other materials acting as substrates for growth of *A. versicolor* are wallpaper glue, wood pulp, rotten straw, stored grain, mouldy hay and fruit juices, jam, meat products and cheese. Growth on the food products mentioned indicates the xerophilic feature of the fungus, as these products have a low water activity (available water).

The fungus is also responsible for biodegradation of military equipment and optical instruments in the tropics.

Specific metabolites

The volatile metabolite geosmin – an undecan – has been demonstrated in connection with growth of *A. versicolor*. It has a characteristic musty, earthy odour, often connected with mouldy houses and is the cause of mucosal irritation such as eye, nose and throat irritation.

Other important metabolites are averufin, cyclopenin, cyclopenol and versicolorin, which have strong antibacterial activity and also specific antifungal properties, and sterigmatocystin, which is toxic and carcinogenic. This compound can be found in many spoiled foodstuffs.

Aureobasidium pullulans

From Latin *aureus*: golden, *pullus*: black-brown.
Synonyms: *Pullularia pullulans, Dematium nigrum, Torula olea, Hormodendrum dermatididis, Cryptococcus meta-niger, Oospora variabilis, Monilia nigra* – and several others.

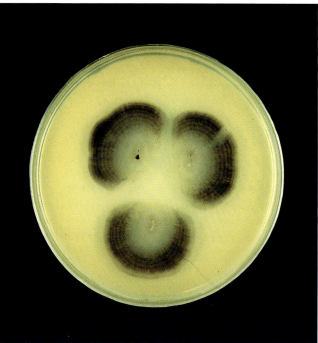

Description

This is one of the most interesting species of the black yeast-like fungi. In culture it exhibits extreme variability in shape and colour. When grown on MEA at 25°C it develops slimy white to cream-coloured colonies or they can be yellow, pink or brown. Later they become more or less black. Colonies will extend to 6 cm in diameter in 7 days. The mycelium is mostly immersed and covered by hyaline, smooth-walled, ovoid, one-celled conidia around 5-7 μm in size, but this can be variable, as secondary smaller conidia are often produced. With age the mycelium will show a marked pigmentation caused by melanin.

Ecology

A. pullulans is a saprophyte with a worldwide distribution. It has also been reported as an endophyte and is known as a primary invader of leaves such as oak, beech, hornbeam, maple and poplar. *A. pullulans* is frequently isolated from soil.

Growth temperature ranges from 2 to 35°C with the optimum at 25°C. During summer the spores are deposited on the leaf surface without attacking the cells. In autumn when the leaves reach senescence, the fungus begins to produce pectinases, which break down the pectic components of the middle layer of the cell walls. This initial damage facilitates the decomposing activity of secondary invaders such as *Cladosporium herbarum*, which attacks the cellulose components of the cell wall.

Practical application

The ability of this fungus to produce decomposing enzymes is utilised in practice in the removal of the unwanted components of raw textile material from plants such as flax or hemp which undergo dew- or wet-retting as a natural manufacturing process.

A. pullulans produces a polysaccharide, pullulan, which is a biodegradable material used for packaging of food and drugs. It can be processed into fibres which have a shiny gloss like rayon and a tensile strength of the same order as nylon. Mixed with other natural fibres, it can be used to make special kinds of paper.

Damaging effects

Agricultural aspects: Problems may occur with contamination of grain such as barley, oats and wheat and of soft fruit such as cherries with lesions, pears which undergo storage or transit, strawberries and oranges. Storage rot of tomatoes may also be a problem.

Building aspects: Indoors *A. pullulans* is often found in connection with dampness, e.g. on damp materials in kitchens, bathrooms and on wet window frames. It often occurs in the silicon kit used for sealing bathtubs and shower screens. On wooden, painted surfaces growth on the wood is able to penetrate the paint, resulting in dark spots. *A. pullulans* is a problem on weathered wood, as it may grow under paint and discolour the wood. Also, it may be resistant to a variety of fungicides used in paint. Growth in paint and in cutting oil is also reported.

Medical aspects: Allergy to *Aureobasidium* is frequently recorded among atopic patients as a positive skin prick test. Whether it is of any significance for the clinical condition of the patients is unclear.

Infections are seldom recorded, but it has been isolated from skin and nails. A very rare case of repeated isolations from blood of a patient with acute leukaemia has been reported.

Specific metabolites

These are not known, but several polysaccharides are produced, e.g. pullulan mentioned above.

Botrytis cinerea

From Greek *botrys*: bunch of grapes,
cinereus: ash-grey.

Description

The colonies are more or less effuse,
grey to greyish brown, often with black
sclerotia. The colonies reach around 10
cm in 7 days when grown on MEA at
25°C, and thus it is one of the very fast
growing moulds.

Seen under a low power dissection
microscope, the colonies, with their
stout brown conidiophores and glisten-
ing heads of pale conidia, look like flo-
wering cherry trees. The conidia are
one-celled, smooth, colourless or pale
brown, spherical to ovoid, 8-11 μm in
size.

Ecology

This cosmopolitan "grey mould" is isolated from many different types of soil
and from a variety of plants and fruit,
where it can act as a damaging plant
parasite.

Minimum temperature for growth:
2°C, optimum: 22-25°C, maximum:
33°C. When grown in the laboratory the
culture often has a characteristic musty
or earthy odour.

The conidiophores are sensitive to
changing humidity, and the liberation of
spores is due to hygroscopic movement
of the spore-bearing organs.

deaux and the Rhineland and is called *Edelfäule,* and in French: *la pourriture noble.*

Damaging effects

Agricultural aspects: The pectolytic activity of *B. cinerea* together with the ability to decompose glucose, cellulose and cutin makes this mould a potent spoilage organism, especially in connection with soft fruits. Within few hours, *B. cinerea* is able to destroy a whole basket of strawberries. Another species, *B. aclada,* is known as the grey mould of onions and is capable of damaging shiploads of onions, legumes or fruit which are handled in world trade nowadays.

B. cinerea has been isolated from deposits at the bottom of water reservoirs. In this habitat it produces benzyl cyanide, which has a special grassy odour.

Medical aspects: Allergy to *Botrytis* is well known. Severe attacks of allergic asthma may occur after indoor exposure, e.g. in greenhouses.

Specific metabolites

Botrydial and botryllin (related to alternariol) are produced, the latter by *B. aclada* (= *B. allii*), which has caused spoilage of onions. It is not known whether the metabolites have any toxic effects.

Practical application

B. cinerea is a very good producer of pectolytic enzymes, which are active at a low pH. Addition of these enzymes into the industrial processing of juice and marmalade facilitates the extraction of juice from fruit pulp due to degradation of the natural pectic substances present in the pulp.

"The noble rot", a positive result of infection with *Botrytis* on grapes, causes the evaporation of water from the grapes with the subsequent increased concentration of sugar. This feature is exploited in the production of wines from Bor-

Candida albicans

From Latin *candidus*: white, *albus*: white.
Synonym: *Monilia albicans.*

Description

On Sabouraud agar, pigmented compact creamy colonies have a slimy yeast-like appearance and will reach 5 cm in diameter within 4 days when grown at 37°C.

The spores are formed by multipolar budding and occur as globose, ovoid, cylindrical or elongate one-celled propagules 3-24 μm in size. Pseudomycelium and thick-walled chlamydospores may often be formed.

Identification of the species is not based on morphology, but as is the case within the true yeasts, based on biochemical tests similar to bacterial identification. The commercially available Nickerson's medium is a useful indicative medium, as *C. albicans* will appear as brown colonies in few days.

Ecology

C. albicans has no natural occurrence in nature but can be isolated from some warm-blooded animals, including human beings. Optimal temperature for growth is 37°C. The upper airways and the intestinal tract are the most important reservoirs for *Candida* in normal healthy humans. The genital mucosa, especially the vagina, is also a natural habitat for *C. albicans.* In most of these cases the fungus is present as a harmless saprophyte. Demonstration or isolation of *Candida* from these foci may often be normal and not a proof of any pathogenic condition.

Recently published theories on *Candida* as an almost ubiquitous villain responsible for every imaginable bad condition in the human well-being lack any scientific background and are more or less ignorant, cynical speculation for commercial reasons.

Practical application

C. albicans is not used in any industrial process because of its role as a potential pathogen. Another *Candida* species, the so-called food-yeast *C. utilis,* has been cultivated on a large scale since the 1940s for animal fodder and to some extent for human food, as it contains large amounts of protein and B vitamins, which are both of high nutritive value. This biomass production is performed by fermentation of sources such as sulphite pulp from processing paper in paper mills.

Damaging effects

C. albicans may infect the skin and nails, the mucous membrane of the mouth and vagina. Humid skin and mucous membranes are the most common sites of infection, but also internal organs may be affected. This can be the lung, the intestinal and urinary tract. Infection of the heart and brain may sometimes occur.

A pathological condition due to *Candida* will occur as a correlation between the fungus and the host. Predisposing factors to *Candida* infection are diabetes mellitus or lowered resistance, either systemic or local. Such conditions can be seen under natural circumstances such

as pregnancy, in newborn children and during senescence. Use of certain types of antibiotic and contraceptive pills predispose to *Candida* infections.

Candida has gained increasing importance as a pathogenic organism, especially in patients with a lowered resistance to infections. This could be the immunocompromised patients, i.e. patients with their immune system either temporarily or permanently partially or completely nonfunctional.

This is seen in patients undergoing chemotherapy as in cancer patients, or it could be due to viral destruction of some of the immunocompetent cells, as seen in AIDS patients.

Infections with *C. albicans* are regarded as being derived from a focus in the patients themselves. Very few *Candida* infections are transmitted and the fungus is not regarded as being contagious.

Specific metabolites

Not known.

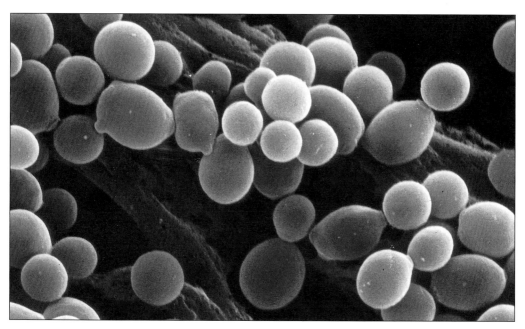

Cladosporium herbarum

From Greek *klados*: a branch, spores in branched chains;
from Latin *herbarum*: of plants.
Synonyms: *Hormodendron, Hormodendrum.*

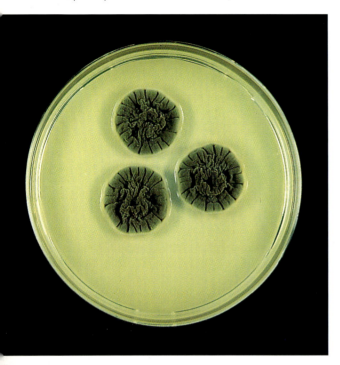

Description

Colonies will reach a diameter of 3-3.7 cm in 10 days when grown on MEA at 18-20°C. The colonies are low, dense, powdery or velvety olivaceous-green to olivaceous-brown. The reverse are greenish-black. The erect, pigmented conidiophores have tree-like branchings, pale to mid-olivaceous-brown or brown and with smooth walls.

Conidia often occur in long, branched chains and show considerable variations in size and septation. Two characteristic forms of spores are produced in *C. herbarum*: (a) lemon-shaped conidia 3-5 μm, and (b) torpedo-shaped, often with two or more cells, e.g. 3-23 μm in size. Both types of spores are golden brown, conspicuously verruculose, sometimes with protuberant scars.

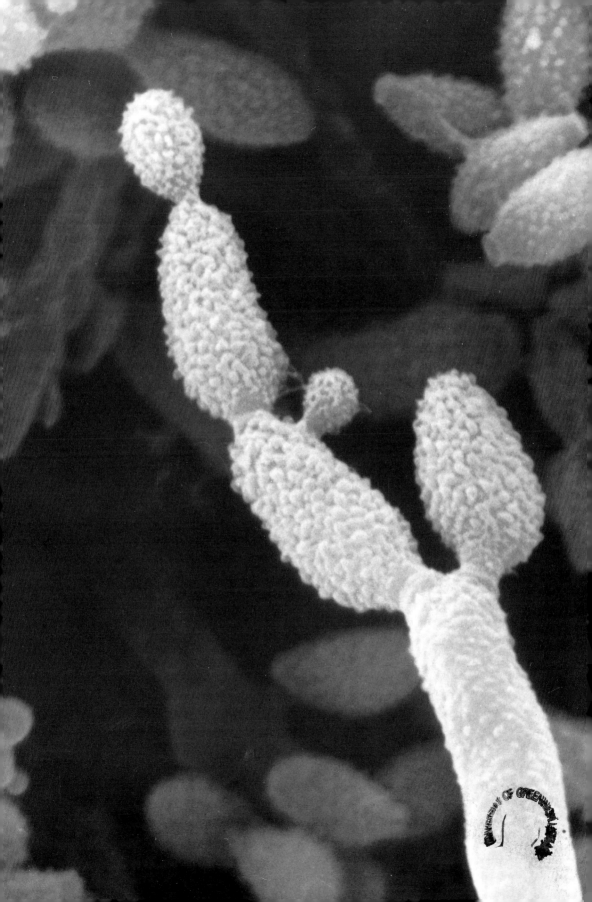

Ecology

The genus *Cladosporium* comprises several very common species, of which conidia from *C. herbarum* are the most frequently encountered in outdoor air in temperate climates. The dry conidia, borne in very fragile chains, easily become airborne and are transported over long distances.

Optimum temperature ranges from 18-28°C, but growth is possible down to – 6°C.

Systematic recordings and mappings which have been done in the Western Hemisphere for years show that *Cladosporium* dominates the mould spore content of air in the late summer and autumn. Spore counts up to 50,000/m^3 of air have been reported in the peak seasons. Half of the mould spore counts made in forested areas consist of *C. herbarum,* and this can be accounted for by heavy sporulation, easy dispersal and buoyant spores.

Indoor records of moulds from houses situated near deciduous forests often show high concentrations of *C. herbarum* colonies, especially in the high season, which is late summer and autumn, indicating that the spores will penetrate the houses from the outdoor environment. High concentrations of *Cladosporium* in the low season, which is winter and early spring, almost always indicate moisture problems.

Outdoors, this mould is isolated from many different types of soil. It is closely connected with plant litter, as the enzymes of the mould are especially suited for attacking cellulose, pectin and lignin, the major components of plant litter.

Practical application

Enzymes produced by *C. herbarum* have been used in the synthesis and transformation of steroid intermediates such as pregnenolone and progesterone, biologically important hormones used in the industrial production of contraceptive pills.

Large-scale cultivation of *C. herbarum* is carried out for the production of biomass used as source material in the pharmaceutical industry for standardised allergen extracts for the specific diagnosis and treatment of airway allergy against this mould.

Damaging effects

The ability of *Cladosporium* to invade many different ecological niches rapidly makes its presence almost ubiquitous and therefore sometimes problematic. Growth is frequently seen on different foodstuffs, e.g. meat in cold stores and butter.

Household and building aspects: As the minimum temperature requirement for growth is below 0°C, *Cladosporium* is often encountered in dirty refrigerators, especially in the reservoir where condensation is collected. On moist

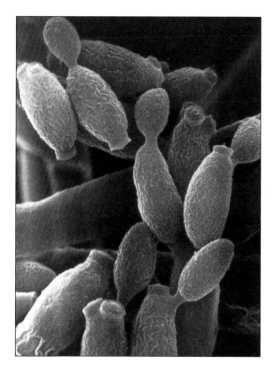

window frames, it can be seen covering the whole painted area with a velvety olive-green layer. Visible growth can be observed even from tight wooden frames around thermo-glass, presumably due to dampness in the room and formation of condensation.

A special category of damage is soiling, where the main cause of discolouring of infected material is the presence of the coloured mycelium as well as the spores. Interior paint work, paper or textiles stored under humid conditions are often discoloured by *Cladosporium*.

Houses with poor ventilation, new as well as older ones, houses with a thatched straw roof and houses situated in low, damp environments may have heavy concentrations of *Cladosporium* spores. These are expressed in high plate counts following domestic mould analyses.

Other damaging effects: *C.* (*Hormoconis* or *Amorphotheca*) *resinae* is frequently isolated from the fuel tanks of aircraft, where the fungus can grow in the water phase of the fuel oil. The oil serves as a source of nutrition, and accumulation of fungal material results. The species also often occurs in kerosene and diesel.

Medical aspects: The above mentioned ability to sporulate heavily, easy dispersal and buoyant spores makes *C. herbarum* the most important fungal airway allergen, which together with *Alternaria,* cause asthma and hay fever in the Western Hemisphere.

Infection is rare. In tropical areas a fungal infection in a skin lesion named chromoblastomycosis may occur. Another common species, *C. sphaerospermum,* can be a parasite of a small octopus (*Eledone*).

Specific metabolites

None known.

Emericella nidulans

(anamorph: *Aspergillus nidulans*)
From Latin: *nidulus*: little nest.

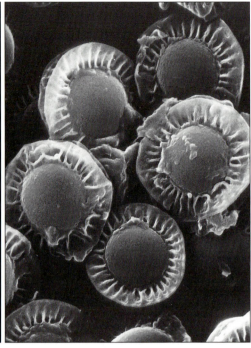

Description

Colonies are velvety, fast growing – 6-7 cm in 7 days on MEA – with numerous green conidia, dull yellow to buff ascomata and off-white Hülle cells. The reverse side of the colony is yellow brown to magenta red.

Sporulation is often heavy, occasionally with few ascomata. Stipes are brown with metulae and phialides. Conidia are globose and rough, 3-4 μm. The ascospores are oblate, red to purple and have two entire crests and smooth convex surfaces. The Hülle cells are globose, 12-24 μm.

Ecology

It is very common in subtropical and tropical soil, spices and other foods dried directly on the soil. It is commonly airborne and sometimes isolated from house dust samples.

The optimum temperature for formation of ascomata lies between 20 and 30°C, while the optimal temperature for development of conidiophores is 36°C. Germination of conidia may occur between 12 and 37°C.

E. nidulans is a good decomposer of starch and cellulose. It is able to utilise hydrocarbons from fuel oil and grow on PVC.

Practical application

The fungus has been used extensively in genetic experiments, to elucidate the penicillin biosynthetic pathway. It produces much less penicillin than *P. chrysogenum*, but has the advantage of having a perfect stage (teleomorph), making genetic investigations easier.

Damaging effects

E. nidulans has been reported as pathogenic and allergenic. The fungus grows strongly at 37°C and has been found in nasal mucosa of horses and other animals, where it has caused haemorrhages. It has been reported as the cause of aspergilloma (fungal ball) in the lungs of immunodeficient patients. This condition is more often caused by *Aspergillus fumigatus*.

Specific metabolites

This fungus produces the carcinogenic mycotoxin sterigmatocystin and several other specific metabolites including asperthecin, penicillin, cordycepin, pentostatin, asperugin, emerin, emericellin, nidurufin, asperline and echinocandin B.

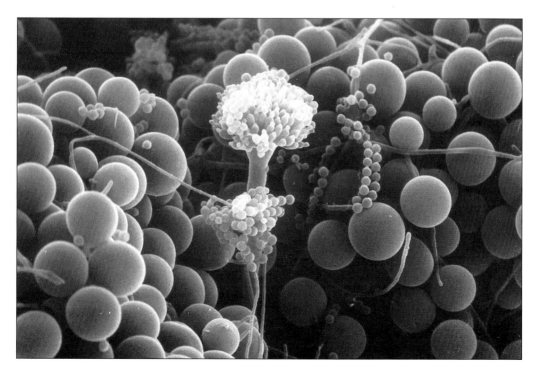

Epicoccum nigrum

From Latin *epi*: over, *coccus*: sphere, *niger*: black.
Synonym: *Epicoccum purpurascens*.

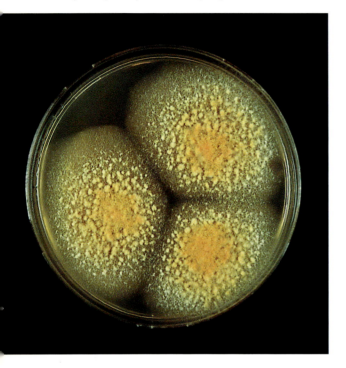

Description

The brightly yellow or reddish coloured colonies reach a diameter of 6-8 cm in 10 days on MEA. Macroscopically, the colonies are very characteristic with a lanose to felty appearance with colours ranging from yellow-orange to bright red or brown, sometimes greenish.

In sporulation, small black spots are developed. The reverse of the culture is similar, but even more intensely coloured. The large, often 15-celled conidia are dark golden brown to dark brown, verrucose, 15-25 μm in diameter, globose to piriform and placed in a fruiting-body, a so-called sporodochi-um. In nature or in culture such sporodochia are visible to the naked eye as pulvinate black spots 100-2000 μm in diameter.

Ecology

Where *Aureobasidium* and *Cladosporium* are the primary invaders in nature in the degradation of dead plant material *Epicoccum* is an extremely common secondary invader of many different sorts of plants. The conidia are actively liberated from the sporodochium by means of hygroscopic movements. In dry weather a whole sporodochium can often release all its conidia simultaneously.

It is found in different types of soil and frequently isolated from air, sometimes from house dust.

The growth temperature ranges from -3°C up to 45°C. The optimum temperature is from 23 to 28°C. For germination, a RH of at least 92% is needed.

When cultivated in the laboratory, the cultures often fail to produce conidia. Sporulation can, however, be induced if the culture is exposed to near-UV (black) light and cultivated on a poor nutrient medium.

The brightly coloured pigments produced in the mycelial part of the fungus consist of components such as β-carotene, torularhodin and rhodoxanthin.

Damaging effects

Agricultural aspects: *Epicoccum* infects seeds from crops such as barley, oats, wheat and corn. Discoloration of moul-dy paper is often caused by *E. nigrum*.

Medical aspects: Airway allergy is reported, but not with a high frequency. Growth at 37°C is possible due to the thermotolerant properties of the fungus. This enables *Epicoccum* to act as a human pathogenic organism, which may cause infection of the skin.

Specific metabolites

The antibiotic substances flavipin, epicorazine A & B, indole-3-acetonitrile and several dyes are produced by this fungus. Furthermore, phenyalanine anhydride, ferricrocin, coprogen, trioricin and an α-naphtopyran are identified as metabolic products.

Eurotium chevalieri

(anamorph: *Aspergillus equitis*)
From Greek *eurotiao:* to decay or become mouldy;
named after F. Chevalier, a French mycologist.

Description

Colonies on CY20S reach 4.5-6.5 cm in one week with vividly yellow ascomata and dull green conidia. Colony reverse is yellow to light orange brown.

Conidia are rough, 3.5-6 × 2.5-4.4 μm and ascospores are oblate, pulley-formed, 4.5-5.5 × 3.5-4 μm and smooth walled with two longitudinal flanges. Smooth-walled stipes and only phialides are present.

Ecology

E. chevalieri is a xerophile (osmotolerant) and grows well on low water activity media and is of worldwide occurrence. It has been found on jam, cakes, leather, cotton, seeds, dried fruit, nuts, spices and cereals. Optimum temperature for growth is 30-35°C with a maximum of 37°C. Tolerance of ascospores of a 10-minute treatment at 80°C is reported.

Cellulose or wood is not attacked. Lipase production is seen on coconut oil emulsion.

When *E. chevalieri* is grown on rice infected with *Aspergillus parasiticus*, the production of aflatoxin from the latter mould is very much reduced.

Practical application

None.

Damaging effects

E. chevalieri is a widespread biodeterioration fungus, spoiling humid leather
and other fabrics.

Specific metabolites

E. chevalieri produce questin, questinol,
flavoglaucin, echinulin, preechinulin,
physcion, erythoglaucin, catenarin,
rubrocristin and viocristin.

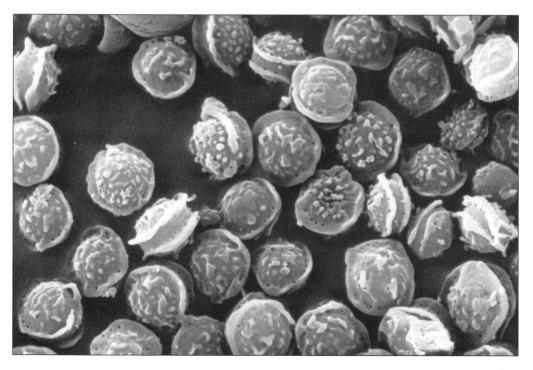

Eurotium rubrum

(anamorph: *Aspergillus rubrobrunneus*)
(formerly: *Aspergillus glaucus* group)
From Latin *ruber*: red.
Synonyms: *Eurotium umbrosum, Aspergillus umbrosus, A. sejunctus, A. ruber.*

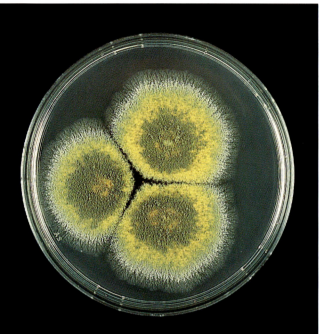

Description

Colonies on CY20S are 3.5-6 cm after one week at 25°C. The fungus has grey-green conidia and yellow ascomata. Orange to red mycelium is present. The reverse side is bright yellow to reddish orange to blackish brown in age.

Conidia are ellipsoidal to apiculate, rough, 4.5-8 × 5-7.5 μm and asco-spores oblate, 4.5-5.5 × 3.5-4 μm, smooth walled with a furrow. Smooth-walled stipes and only phialides are present.

Ecology

This fungus is of worldwide distribution and has been isolated from stored mouldy hay in Iceland and frequently from Finnish barns, which is quite natural, as it is known as an early colonist of decomposing straw.

Optimum temperature for formation of conidia and ascomata is 25-30°C. Maximum for growth is 37-40°C.

Like *Eurotium chevalieri*, it is a xerophilic fungus and it is found in many of the same habitats as this fungus.

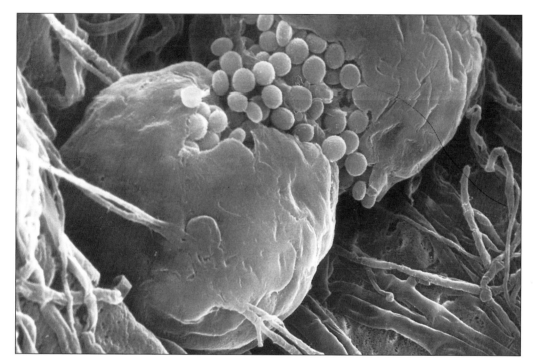

Practical application

E. rubrum has been reported to be used for fermentation of fish in Japan.

Damaging effects

From Finland *E. rubrum* has been reported, under the name *Aspergillus umbrosus*, to cause farmer's lung (extrinsic allergic alveolitis) among farmers.

It is also a widespread biodeteriogen and destroys material and food such as leather and cotton, spices and meat and fermented and dried food products.

It is recorded from silage, from different crops such as corn, where it is found to be the most important invader if the corn is stored at moisture contents between 14 and 23%.

Specific metabolites

Among the specific metabolites reported from this species are: echinulins (incl. preechinulin and neoechinulin), flavoglaucins (incl. dihydroauroglaucin, isodihydroauroglaucin and tetrahydroauroglaucin) and anthraquinones incl. emodin, questin, questinol, catenarin, physcion, erythroglaucin, physcionanthron-dimer and physcion anthron A & B.

Fusarium graminearum

(teleomorph: *Gibberella zeae*)
From Latin *fusus*: a spindle, the shape of the spores, *graminis*: grass.
Synonym: *F. roseum*.

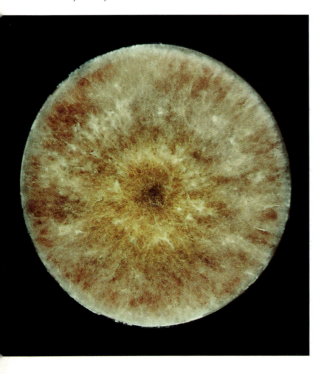

Description

Colony on OA at 25°C is fast growing and has a typical floccose, greyish rose to red, often vinaceous red appearance. Colonies will reach a diameter of 9 cm in 4 days. Aerial mycelium whitish, becoming brownish to rose. In contrast to other species of *Fusarium*, it is not easily wetted with water.

Sporulation is often sparse. Macroconidia are gondola-shaped with often 5-6 septae, 4-5.5 × 40-60 μm in size. Microconidia are absent. Chlamydospores are often formed within the macroconidia.

Ecology

This fungus has a worldwide distribution. It is isolated from soil and from plants belonging to the grass family, where it can also act as a parasite. It attacks seeds of cereals, in particular barley and corn. Massive growth in water-damaged carpets in a school is reported. Optimum temperature for growth is 25°C; toxin production is optimum at lower temperatures.

112

Practical application

The most important specific metabolite produced by this fungus (zearalenone) is a hormone-like compound. When used in very small doses, it is regarded as applicable in oral contraception for humans and as an anabolic steroid at levels lower than 1 ppm. A derivative of the substance has been patented as a growth stimulant in animals.

Clinical trials are in progress for investigating the possible positive effects in treatment of the postmenopausal syndrome in women.

F. graminearum is used for the production of quorn (mycoprotein) for human consumption. The product contains many fibres and is used in pies in the United Kingdom etc.

Damaging effects

The most harmful feature of this fungus is the toxin production during storage of an infected plant or crop. Infection of e.g. corn is favoured by cool, wet conditions. The subsequent production of the specific metabolite zearalenone in the stored corn is enhanced by high moisture and alternating moderate and low temperatures. The damaging effects will be described below.

Specific metabolites

The most important specific metabolite produced is the oestrogenic compound zearalenone, also known as F-2 toxin, and the type B trichothecenes: deoxynivalenol (= vomitoxin), 1-acetyldeoxynivalenol, fusarenone X and nivalenol. Zearalenone is not acutely toxic and can be regarded as a nonsteroidal fungal hormone. Interestingly, ingestion of controlled amounts, as mentioned above, have different positive effects, while larger amounts give rise to clinical manifestations known as the oestrogenic syndrome. The clinical symptoms of the toxic influence are very significant in pigs, which are the most sensitive animals to this toxin, but cattle are also affected. Ingestion of infected corn affects the reproductive organs of the animals.

The possible effects in human beings of exposure to zearalenone after ingestion and especially after inhalation are not known. It has been a subject of speculation as to whether zearalenone could induce hormone-dependent tumours in women.

Other specific metabolites are sambucoin, 4-acetamido-2-butenoic acid, aurofusarins, butenolide, culmonin and fusarin C.

Fusarin C

Zearalenone

Deoxynivalenol

Fusarium oxysporum

(teleomorph: *Nectria haematococca* var. *brevicora*)
From Latin *oxysporum:* pointed spore.
Syn: *F. bulbigenum, F. vasinfectum, F. angustum, F. bostrycoides* and many others.

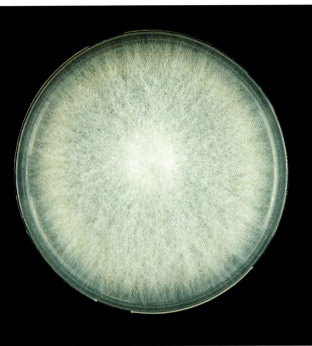
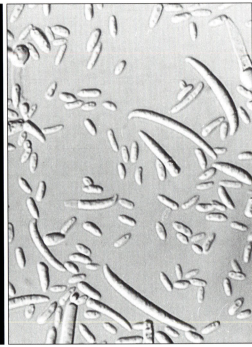

Description

This fungus is one of the most variable *Fusarium* species. Although *F. oxysporum* grows relatively fast, the colonies grow more slowly and more sparsely than those of *F. graminearum*, and they will attain a diameter of 4-5 cm in 4 days when grown on OA or PSA at 25°C. The mycelium is at first floccose, becoming felty, whitish or peach usually with a purple tinge, more intense near the medium surface. The reverse side of the culture has purple shades.

Microconidia with 0-2 septae are present. Macroconidia are sparse; in some strains they are 3-5-septate, gondola-shaped and 3-5 × 27-40 μm in size. In many isolates chlamydospores are formed. These resting organs are essential for survival in soil.

Ecology

F. oxysporum has a worldwide distribution as a saprophyte, but might also be a parasite on many different plants. It is isolated from cereal grains and from fruit and vegetables.

F. oxysporum possesses a higher competitive saprophytic ability in comparison with other root parasites. Dispersal

116

of the conidia in the soil may be brought about by water movements.

The fungus has an optimum temperature for growth between 25 and 30°C and a growth range from 5 to 37°C. It tolerates a high carbon dioxide concentration and a salt concentration up to 18%.

Organisms antagonistic to *F. oxysporum* are species of *Streptomyces*, *Trichoderma* and *Paecilomyces*.

Practical application

The fungus is used for production of urate oxidase, edible protein substances and for chemical transformations. It is also used for inducing cellulase production and for estimation of cellulase activity.

Damaging effects

F. oxysporum is one of the most economically important species of this genus, as it can be pathogenic to numerous plants. It causes storage rot in cereals such as corn. It is isolated from flax, cotton, groundnuts, soya, peas, onion bulbs, potatoes, bananas, oranges and apples.

It is also the causal agent of damping off in cultivated mushrooms. Rats have been killed after ingestion of maize contaminated with this fungus.

Nail infections (onychomycosis) are reported.

Specific metabolites

Ageing cultures produce inhibitory compounds to other fungi. Most of these inhibitors are volatiles such as aldehydes and alcohols. Nonanoic acid is one of the compounds identified.

Oxysporone, enniatin A + B, lycomarasmin, fusarubin, javanicin, marticin, bostrycoidin and fusaric acid are produced together with bikaverin (=lycopercin) and moniliformin.

Mucor plumbeus

From Latin *mucor*: fungus, *plumbeus*: leaden.
Synonym: *Mucor spinosus*.

Description

A colony grown on MEA at 25°C will cover the Petri dish (9 cm) in 4 days. Dark grey or light olive-grey up to 2 cm tall needle-like sporangiophores with slightly encrusted walls and a terminal sporangium with spinulose walls.

Asexual spores are developed around a piriform columella with typical projections. The spores are globose, sometimes ellipsoidal or irregularly shaped 7-8 µm in diameter, yellowish brown and slightly rough-walled.

Ecology

M. plumbeus has a worldwide distribution. It is isolated from soil in many locations from Murmansk to South Africa and central China with a very wide pH range in the soil.

It is frequently found in hay, stored seeds, horse dung and almost always in house dust regardless of geographical location. Sewage sludge, children's sandpits and feathers from free-living birds are other growth foci. *M. plumbeus* is frequently found in air samples.

Growth temperature ranges between 5 and 35°C with optimum between 10 and 25°C. Sporulation is poor between 28 and 30°C and maximum temperature for growth is 37°C. Another factor which

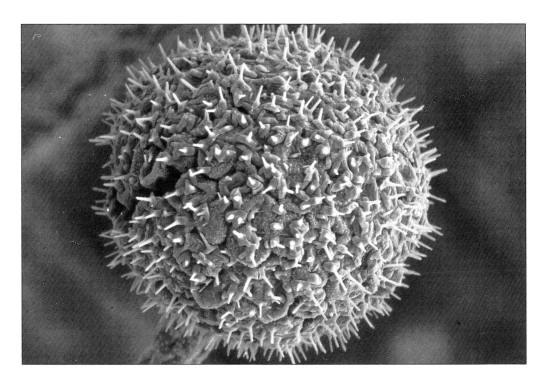

suppresses sporulation is the secondary metabolite rubratoxin B produced by *Penicillium crateriforme*. Garlic extract is a strong inhibitor for growth of the fungus in vitro.

Practical application

Although the fungus possesses strong pectolytic and chitinolytic enzymatic activity, it has never been used for industrial production of enzymes.

Another closely related species, *M. racemosus*, produces the enzyme diastase, which can be used for conversion of starch to sugar.

The similar but thermophilic species *Rhizomucor miehei* is industrially exploited for the ability to produce proteases, which are used as rennet protein in cheese production. Other Mucoraceae are active agents in the so-called starters used for initiating alcoholic fermentation in Asian countries.

Damaging effects

M. plumbeus is often seen in domestic households, where it attacks different sorts of left-overs, soft fruits, juice and marmalade.

Building aspects: It is one of the most

common moulds isolated from house dust. It is found in old dirty carpets and in indoor air. Accumulated dust in ventilation ducts, especially exhaust ducts, may contain high concentrations of viable *Mucor* spores.

Medical aspects: Heavy exposure to *Mucor* spores may cause specific airway allergy. Wood chips as well as sawdust are often attacked by *M. plumbeus*, and handling this material may lead to Type III allergy (EAA). These wood chips are used for heating, while the sawdust is applied in drying mink furs before they are processed into fur coats. Inhalation of spores of this *Mucor* species and other fungi originating from these substrates may cause "wood chips disease" and "furrier's lung".

EAA is also reported in connection with exposure to *Mucor* from ducts in a forced air ventilation system. Asthmatic reactions to *Mucor* (Type I) are described. *Mucor* is an opportunistic pathogenic organism which attacks individuals with lowered resistance (immunocompromised patients). Infections especially in domestic animals are recorded, e.g. mastitis in cows.

Specific metabolites

Mycotoxins are not described, but old cultures produce a substance which inhibits further growth of the fungus and promotes autolysis of the culture.

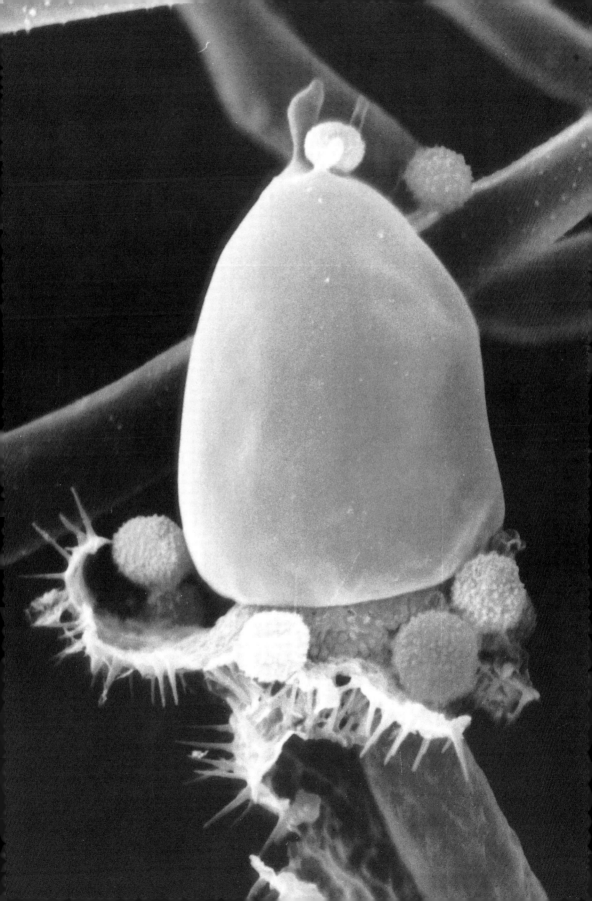

Paecilomyces variotii

From Greek *poikilo*: varied.
Species name from Dr. Variot, a French medical doctor.
Synonym: *Spicaria divaricata*.

Description

Colonies on MEA or V-8 agar grow rapidly, reaching a diameter of 3-5 cm within 7 days at 25°C, consisting of a dense felt with a powdery appearance, a yellow-brown or sand colour and with a sweet aromatic smell.

The conidiophores look to a certain degree like *Penicillium* conidiophores, but phialides taper to a long cylindrical neck.

Conidia are one-celled, hyaline to yellow and smooth-walled. They vary in shape and size, 3-5 × 2-4 μm in long chains of subglobose, ellipsoidal or fusiform cells. Chlamydospores are often seen.

Ecology

This mould, which is frequently isolated from air and dust, is thermotolerant to thermophilic and often found in warm climates and arid regions.

Growth optimum ranges from 25 to 35°C with maximum at 55°C. It is therefore often isolated from decomposing plant material which develops heat such as decomposing straw, garden compost and other self-heating substrates.

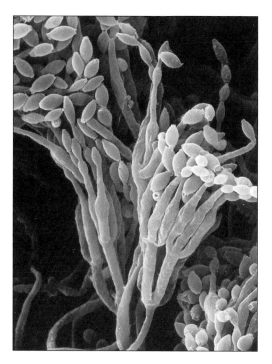

Practical application

Under the name Pekilo, *P. variotii* was used for single cell protein produced by a Finnish factory and used for feed for chicken. Whether it is still produced is uncertain.

Damaging effects

It is a common agent in the deterioration of different foodstuffs; e.g. fruit juice undergoing pasteurisation can be attacked. Because of the production of thick-walled chlamydospores and mycelium, certain strains can be heat-resistant.

Cereals in airtight storage can be dominated by *P. variotii*. It is isolated from different legumes and from cottonseeds as well as from jute fibres and paper. It attacks PVC and timber and gives infected oak wood a yellow discoloration. Examples of other unusual substrates are optical lenses, leather, photographic paper and tobacco for cigars. Recently, a resistance to traditional food preservatives such as sorbic, benzoic and propionic acid has been reported.

Medical aspects: *P. variotii* and other species of *Paecilomyces* have been reported as causative agents of allergic alveolitis and humidifier disease. It has also been demonstrated as causing mycoses in dogs, cattle and other animals. In humans it has been isolated from a heart inflammation and from lachrymal sacs. As an opportunistic pathogen it should be handled with care.

Specific metabolites

Ferrirubin, viriditoxin, indole-3-acetic acid, fusigen, which has an antifungal effect, variotin and a penicillin-like substance have been detected. Sometimes patulin may be formed by *Paecilomyces* strains.

Penicillium aurantiogriseum

From Latin *penicillus*: an artist's brush,
aurantiogriseum: golden yellow grey.
This species represents a complex of penicillia which are also known under
names such as *P. cyclopium, P. viridicatum* and *P. polonicum*.

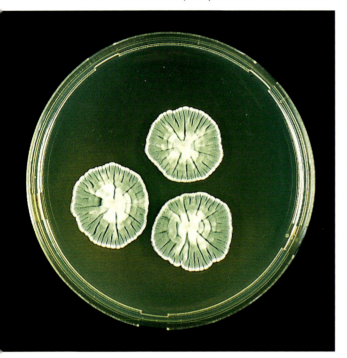

Description

Colonies on CYA reaching a diameter of
1.8-3.0 cm in 7 days and on YES agar
3.6-5.2 cm.

The conidia are grey green, and the
colony reverse is creamish yellow to
yellow brown on CYA and brightly yellow on YES agar.

The conidia are broadly ellipsoidal:
3.5-4.2 µm × 2.8-3.4 µm. The conidiophore stipe is rough walled and
branched twice, i.e. one closely appressed, asymmetrically placed side
branch. These two branches each have
several side branches (= metulae).

Ecology

P. aurantiogriseum is of worldwide occurrence in cereals and air.

Optimum temperature for growth is
25°C, but it grows well from 0°C to
33°C.

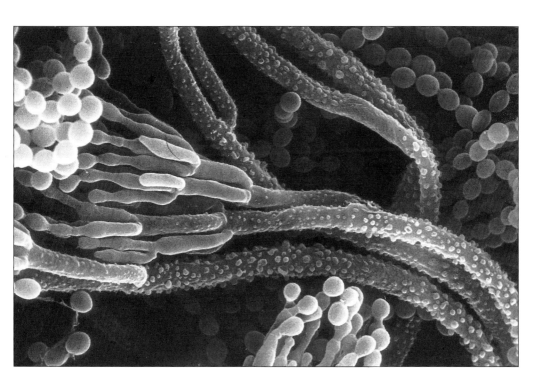

Practical application

None.

Damaging effects

Isolates of this species are often found in materials and fabrics undergoing degradation.

Specific metabolites

Species of the *P. aurantiogriseum* complex are particularly common in the Balkan countries and have been mentioned as possible agents in the Balkan endemic nephropathy. They can pro-duce some nephrotoxic glycopeptides, causing karyomegaly in the kidneys, and also the neurotoxin verrucosidin together with penicillic acid.

It produces a series of terpenoid volatiles of unknown health implications. Usually these moulds also produce an unpleasant smell suggesting soil.

Penicillium camemberti

From the city Camembert in France.
Synonyms: *P. candidum, P. album.*
Domesticated form of *Penicillium commune.*

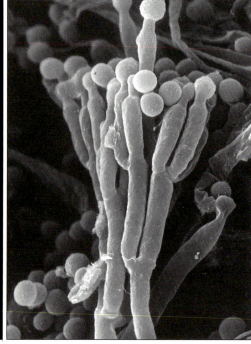

Description

Colonies are rather slow growing, 2.0-3.5 cm in 7 days, white, raised with poor sporulation. Conidia are hyaline or grey green. Colony reverse is cream yellow. Mycelium appears white but may turn yellow or pink with age.

Conidiophore stipes are smooth and rough, asymmetrically two stage branched, with large smooth subglobose to broadly ellipsoidal conidia, 3.5-5.0 × 3.2-4.2 μm.

Ecology

P. camemberti has only been found on white fermented soft cheeses or on hard cheeses which were contaminated by this mould in the store. Its wild type, *P. commune*, has been found on wood, in indoor air and as a common contaminant on cheese, meat and nuts. It grows well on creatine sucrose agar. The fungus grows optimally at 18-28°C.

Practical application

P. camemberti is used for production of white cheeses, Brie, Camembert etc.

126

Damaging effects

Other types of cheese can be contaminated by this mould in the store.

Occupational allergy in connection with production of soft white cheeses, due to inhalation of conidia liberated from the surface of the cheeses has occasionally been reported.

Specific metabolites

P. commune produces the specific metabolites cyclopiazonic acid, cyclopaldic acid, palitantin, fumigaclavine A & B and rugulovasine A & B. Some of these metabolites, notably cyclopiazon-ic acid, are also produced by *P. camemberti*. It primarily produces the latter mycotoxin after prolonged storage at 25°C and much less or nothing at 8°C. It is therefore recommended that the soft white cheeses be stored in a refrigerator. Furthermore, they should only be placed at room temperature one hour before consumption.

Penicillium chrysogenum

From Latin *chrysos*: golden yellow, *geno*: to cause or form.
Synonyms: *P. notatum, P. griseoroseum, P. meleagrinum, P. cyaneofulvum.*

Description

Colonies on CYA are fast growing, 3.2-4.5 cm after one week at 25°C, and very fast on YES agar, 4.2-6.2 cm. On CYA the colonies are velutinous with a broad white mycelium margin which is raised and with numerous often yellow exudate droplets and a yellow or less commonly creamish reverse. The reverse side on YES is always yellow, and the colonies sporulate strongly.

The odour of the culture is aromatic, fruity or spicy, often suggesting apples or pineapples, but sometimes this odour is not pronounced. The conidia are blue to dull dark green *en masse*.

The conidiophore stipes are smooth and two or more stage branched. The branches are often divergent. The conidia are smooth and subglobose to broadly ellipsoidal, 2.5-4 × 2.3-3.5 μm.

Ecology

This well-known fungus is very common in desert sand, dried foodstuffs, spices, dry cereals and in indoor air and house dust. Optimum temperature for growth is 23°C, with a temperature range from 5-37°C.

P. chrysogenum can serve as food for the common storage mites *Acarus siro* and *Tyrophagus putrescentiae*.

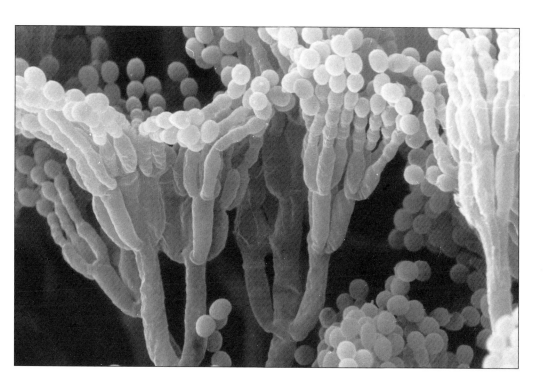

Practical application

Isolates of *P. chrysogenum* (formerly *P. notatum*), which was the first known producer of penicillin, and mutants derived from this are still used for production of penicillin F and G.

Specific metabolites

Apart from producing penicillin, *P. chrysogenum* produces the alkaloids roquefortine C, meleagrin and chrysogine.

Damaging effects

P. chrysogenum is often found in mouldy buildings where it destroys different building material, e.g. wallpaper. It also grows well on the glue on the reverseside of the wallpaper and on moist chipboards and is found in paints.

Penicillium expansum

From Latin *expandere*: to expand.

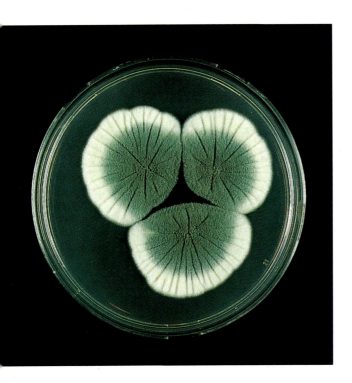

Description

Colonies on CYA fast growing, 2.8-5.0 cm after one week at 25°C, and very fast on YES agar, 3.6-6.2 cm. On CYA the colonies are mealy with a broad white mycelium margin, raised and with numerous often brown exudate droplets and a yellow brown to dark brown or less commonly creamish reverse. The reverse side on YES is always strongly yellow and the colonies sporulate strongly, with a cream to orange mycelium. The conidia are blue to dull green *en masse*.

The culture has a pronounced aromatic-fruity odour suggesting apples.

The conidiophore stipes are smooth and asymmetrically two stage branched. The conidia are smooth and ellipsoidal, 3.0-3.5 × 2.3-3.0 μm.

Ecology

P. expansum produces the typical blue mould rot of apples, forming concentric rings of synnemata (= coremia). It is very common on rotting pomaceous fruit, on nuts and fresh herbs and is often found in indoor air and house dust.

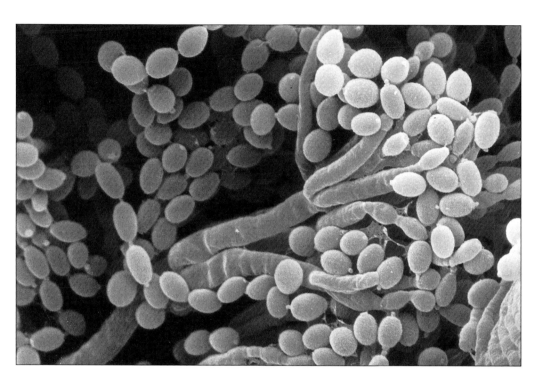

Practical application

P. expansum has been patented for production of animal feed by degradation of white wine vinasse.

Damaging effects

The most severe problem caused by *P. expansum* is the rotting of pomaceous fruits used for juice production. The mycotoxin patulin may contaminate the final product since this compound is not degraded by pasteurisation.

Specific metabolites

P. expansum produces the mycotoxins patulin, which affects killer cell phagocytosis, and citrinin, which is nephrotoxic. Other mycotoxins are chaetoglobosin C and roquefortine C. Furthermore, expansolide is produced.

Penicillium glabrum

From Latin *glaber*: smooth, the surface of the conidia.
Synonym: *P. frequentans*.

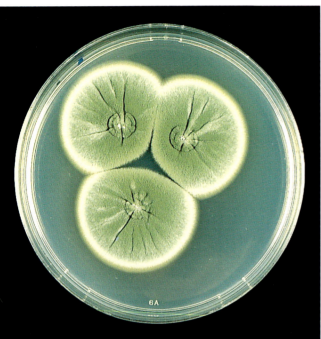

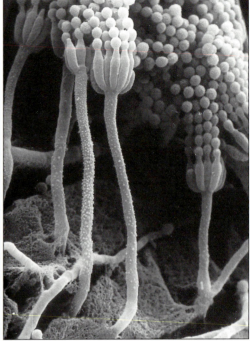

Description

Colony on CYA is fast growing, 3.4-4.4 cm in one week, and has a velutinous surface. Sporulation is very heavy, with dull to dark green conidia. The reverse side of the culture is yellow to reddish brown.

The conidia are globose, smooth to finely roughened, 2.8-3.8 μm, and borne in long columns. The conidiophore stipes are unbranched and short to quite long, 25-150 μm, and smooth.

and has been isolated from all kinds of substrate, especially soil, spices, herbs, compost, cereals, paper, paint and diesel fuel. It is especially common in indoor air of homes and factories.

Optimal temperature for growth is 25°C, but the fungus cannot grow at 37°C.

Ecology

P. glabrum has a worldwide distribution

Practical application

Earlier this species had the name *Citromyces glaber,* referring to the strong production of citric acid. It was originally used for production of this important organic acid, but later it was found that *Aspergillus niger* was a much better producer. It has been patented as a producer of acid proteases and production of animal feed by degradation of white wine vinasse.

Damaging effects

P. glabrum is able to grow in carbonated beverages and on wine corks and can thus deteriorate such beverages. It also attacks onions.

Specific metabolites

Only few metabolites are known from this fungus, the most important being citromycetin, a hepatotoxin.

Penicillium polonicum

From Latin *polonicus*: from Poland.
Synonym: *P. verrucosum* var. *cyclopium pro parte.*

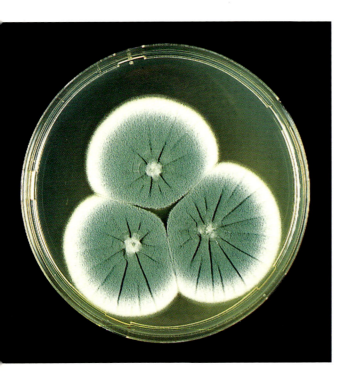

Description

Colonies on CYA grow fast, reaching a diameter of 2.4-3.7 cm in 7 days, and on YES agar 3.6-5.2 cm and are strongly sporulating.

The conidia are blue green and the colony reverse is creamish yellow to yellow brown on CYA and brightly yellow on YES agar.

The conidia are broadly ellipsoidal 3.5-4.2 μm × 2.8-3.4 μm. The conidiophore stipe is rough-walled and branched twice, i.e. one closely appressed asymmetrically placed side branch. These two branches each have several side branches (= metulae).

Ecology

P. polonicum is an old but correct name for a species of worldwide occurrence in cereals, dried meat products and in indoor dust and air. For instance it has been found as the dominating species both in Kenyan and Danish wheat samples. Recently it was isolated from the walls of a mouldy house and appeared to be common indoors.

It grows moderately well on creatine sucrose agar. Optimum temperature for growth is 25°C, but it grows well from 0°C to 33°C.

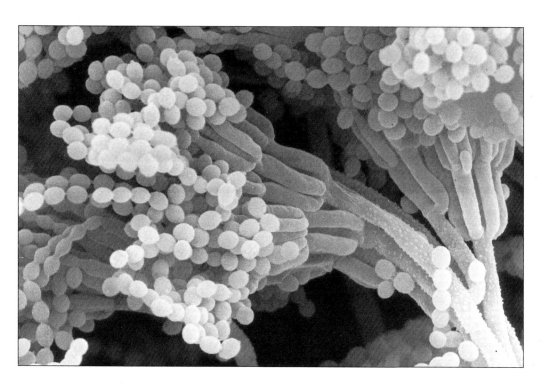

Practical application

None.

Damaging effects

Isolates of this species are often found in materials and fabrics undergoing degradation.

Specific metabolites

P. polonicum is also common in the Balkan countries and has been mentioned as a possible agent in Balkan endemic nephropathy. It produces some nephrotoxic glycopeptides, verrucosidin and penicillic acid. Other specific metabolites of unknown toxicity include 3-methoxyviridicatin, verrucofortine and anacine.

It produces a series of terpenoid volatiles of unknown health implications. Usually this mould also produces an unpleasant smell suggesting soil.

Penicillium roqueforti

From the town of Roquefort, France.
Synonyms: *P. gorgonzolae, P. stilton.*

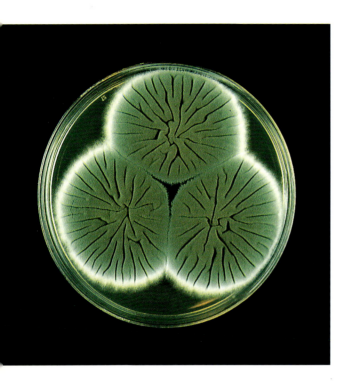

Description

Colonies on CYA are broadly spreading, 3-6.2 cm in one week, velutinous, thin with arachnoid margins and a blackish-green reverse.

Microscopically, this fungus has very rough stipes and smooth large globose conidia (3.5-4.5 μm). It is asymmetrically two stage branched.

Ecology

P. roqueforti has primarily been found in blue cheeses and in other food products, especially rye bread, on blue-stained wood, in forest soil and in silage. It can grow in 0.5% acetic acid, on creatine-sucrose agar, at relatively high concentrations of ethyl alcohol and at low oxygen and rather high carbon dioxide concentrations.

Practical application

This characteristic fungus is used for the production of blue-veined cheeses, such as Gorgonzola, Stilton, Roquefort, Danablue etc. It is also used for aroma-like compounds.

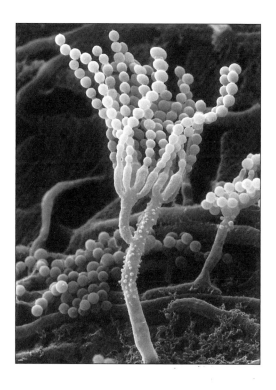

Damaging effects

P. roqueforti is a particular problem on rye bread and other preserved or acid-treated foods (propionic acid, sorbic acid, acetic acid, benzoic acid). It can also grow on or in homemade wine and in corn silage, in which it forms large football-like clumps.

Specific metabolites

The fungus produces roquefortine C, isofumigaclavine A & B, mycophenolic acid, PR-toxin and marcfortins. These metabolites have caused mycotoxicoses in animals (the feedstuff has usually been silage). Although most starter cultures of this fungus used for the fermentation of blue cheese may produce these metabolites in laboratory experiments, the compounds are either unstable or produced in extremely low amounts in blue cheeses.

Rhizopus stolonifer

From Greek *rhiza*: root, *pous*: foot;
from Latin *stolo*: stolon, *fero*: to bear.
Synonym: *Rhizopus nigricans*.

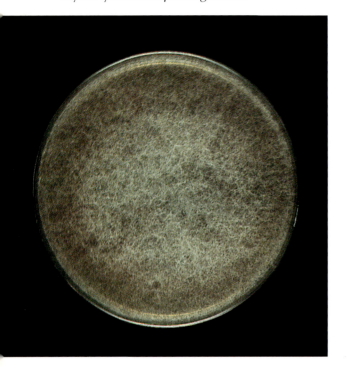

Description

Colonies very fast growing, covering a
9-cm Petri-dish of V-8 agar and other
media in 3 days at 25°C with a reddish
grey-brown mycelium more than 2 cm
high. Easily recognisable by its hyaline
to brown stolons, numerous branched
brown rhizoids and shiny black sporangia.

Spores are mostly oval, pale brown
and ridged, 7-12 × 6-8.5 μm. A typical
microscopic picture shows the flattened
columella, which appears after spore
liberation looking like the cap of a
mushroom.

Ecology

R. stolonifer is one of the most common
members of the group Mucorales. It has
a worldwide distribution.

Growth temperature ranges from 10 to
33°C with the optimum at 25°C.

It is found in soil often from dry areas,
garden compost and municipal waste
and is very frequently isolated from
house dust, wood pulp, dung, honey-
combs and different nuts, fruits and
seeds. Often isolated from forgotten left-
over food or fruit and vegetable garbage.

Degradation of chitin and pectin has
been demonstrated by different *Rhizopus* species. The mycotoxin aflatoxin

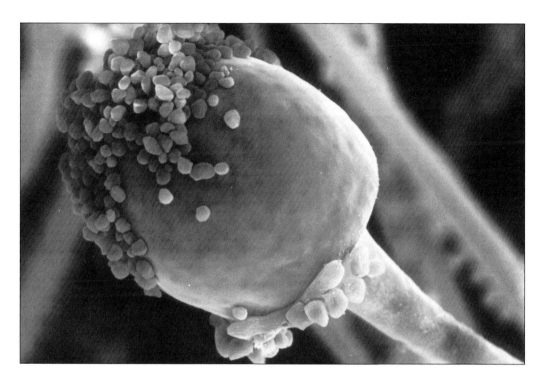

is able to inhibit growth of *Rhizopus*.

Practical application

Enzymes such as proteases, amyloglu-cosidases, lipases and pectinases from different species of *Rhizopus* are used in the food-processing industry, e.g. bread, beer and wine production.

Damaging effects

Rhizopus, together with *Mucor*, is one of the most quickly invading organisms able to contaminate many kinds of stored food product such as wheat, rice, tomato and other vegetables.

Medical aspects: Spores from this fungus become airborne easily. Heavy exposure to *R. stolonifer* has been described from Swedish sawmills causing extrinsic allergic alveolitis, the so-called "sawmill lung".

Processed timber stored at controlled humidity offers excellent growth conditions for this fungus, which is able to cover large surfaces with heavy sporulation and spore emission after a few days' growth.

Positive skin prick test to this mould is also reported. Related species of *Rhizopus* are opportunistic pathogens.

Specific metabolites

Rhizonine.

Scopulariopsis brevicaulis

From Latin *scopula*: little broom;
from Greek *opsis*: like;
from Latin *brevis*: short, *caulis*: stalk.

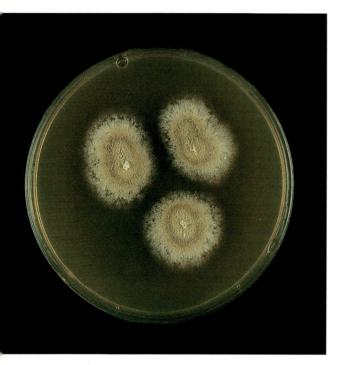

Description

A colony reaches a diameter of around 5 cm in 7 days when grown on MEA at 25°C. The colour is at first light brownish rose, later on dull brown with a powdery appearance, never green coloured.

Conidia are globose to ovoid with a distinctly flat base, verrucose at maturity 5-8 × 5-7 μm in size, rose-brown to brown coloured.

Ecology

S. brevicaulis is an ubiquitous fungus, very common and frequently isolated

from soil, house dust, old carpets, grain, tobacco dust, fruit and nuts, dung, wallpaper and dairy products.

Optimal growth temperature ranges between 24 and 30°C. Maximum is 37°C.

Enzymatic activity from pectinase, amylase, chitinase and proteinase has been demonstrated.

An interesting feature of *Scopulariopsis* is the ability to convert inorganic nitrogen to organic N-compounds in vitro. This includes secretion of a peptide into the medium after complete absorption of the inorganic nitrogen

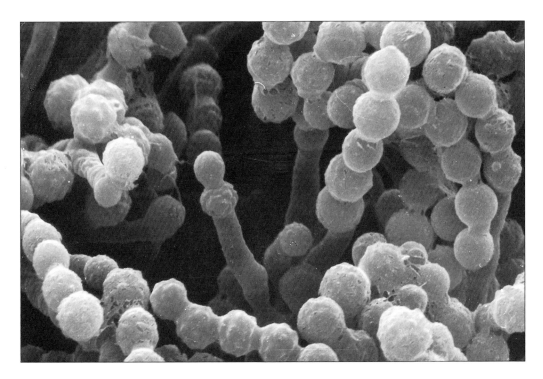

source. Furthermore, it is able to decompose arsenic compounds, which allows it to grow on paint containing arsenic, present on certain types of green wallpaper. Substances containing arsenic compounds are frequently applied to fitted carpets for anti-static purposes, and this may explain why there is often abundant growth of *Scopulariopsis* in such carpets.

Practical application

None.

Damaging effects

Agricultural aspects: As *S. brevicaulis* is able to attack a wide range of organic substrates, it can damage dry fodder and seeds of crops such as wheat and barley. Hay, especially when self-heated, can be attacked, just as organic matter with a high percentage of protein, like milk and cheese, can be decomposed.

Medical aspects: *Scopulariopsis* is a potential pathogenic fungus and may cause infection of the skin and, even more common, infection of the nails. Chronic skin diseases in horses and dogs due to this mould have been reported.

Tobacco leaves under industrial processing as well as tobacco dust can be a substrate for *S. brevicaulis*.

Cases of occupational allergy in this industry are reported, with conidia from this mould as the causal agent. Extrinsic allergic alveolitis has been demonstrated in workers in tobacco factories, e.g. in the Former Yugoslav Republic of Macedonia and in Denmark. Positive skin prick test to the mould is also reported.

Specific metabolites

Ethylene and deacetoxycephalosporin C are produced.

The above-mentioned ability to liberate free arsenic when growing on substrates containing arsenic-compounds has more or less confirmed the theory of Napoleon's death as being caused by arsenic poisoning. Napoleon was forced to spend the last year of his life on the very humid island of St. Helena, situated in the Atlantic Ocean south of the equator. During his stay he had many health problems, including stomach trouble. The walls in his bedroom were covered with a tapestry decorated with green ornaments. Recently fragments of this tapestry have been investigated, and findings showed a growth of *S. brevicaulis* and a substantial amount of arsenic compounds in the green paint. Rather high concentrations of arsenic in Napoleon's hair and nails – structures in which arsenic will accumulate – have led to the conclusion that Napoleon died from inhalation of volatile arsenic emanating from the tapestry in his bedroom on St. Helena.

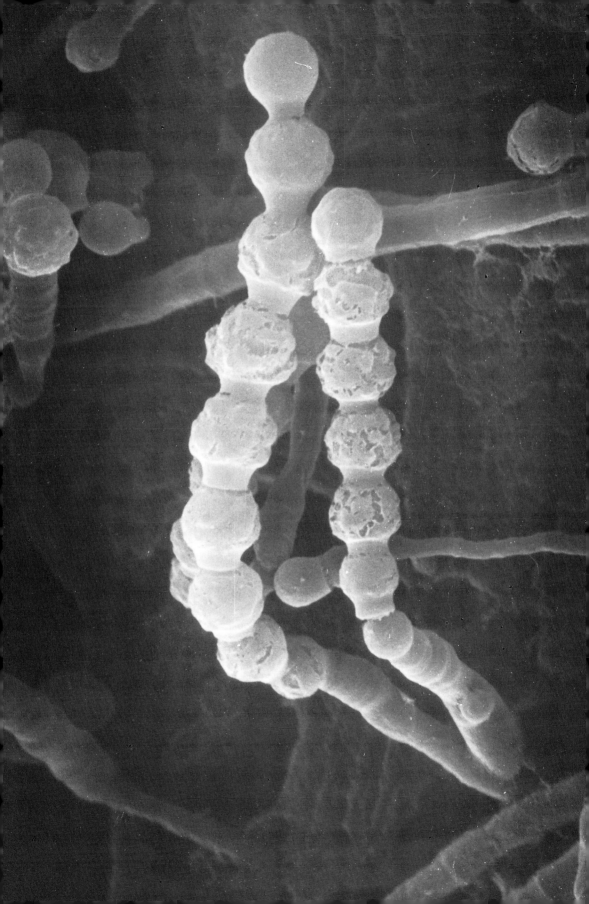

Stachybotrys chartarum

From Greek *stachy*: progeny, *botrys*: bunch of grapes, *charta*: paper.
Synonym: *Stachybotrys atra*.

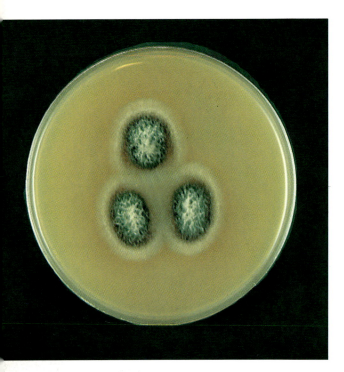

Description

Colonies on MEA rather slow-growing: 2.5-3 cm in 7 days at 25°C with a black to black-green appearance.

Conidiophores are brown, simple or sympodially branched. At the apex 6-10 swollen flask-shaped, light brown phialides are present.

Conidia are ellipsoidal and one-celled, 5-7 × 8-12 μm. At first hyaline and smooth-walled, later on dark and verrucose.

Ecology

Stachybotrys has a worldwide distribution, but grows only on wet substrate. It is strongly cellulolytic and grows from 2 to 40°C with optimum at 23-27°C. It has been found on paper, seeds, soil, textiles and dead plant material.

Only a small percentage of the conidia are viable in laboratory cultures. Furthermore, growth is inhibited by different species of *Penicillium*.

Damaging effects

This mould can be a problem, as it is able to disfigure different materials, including building materials, and because of the production of toxic metabolites.

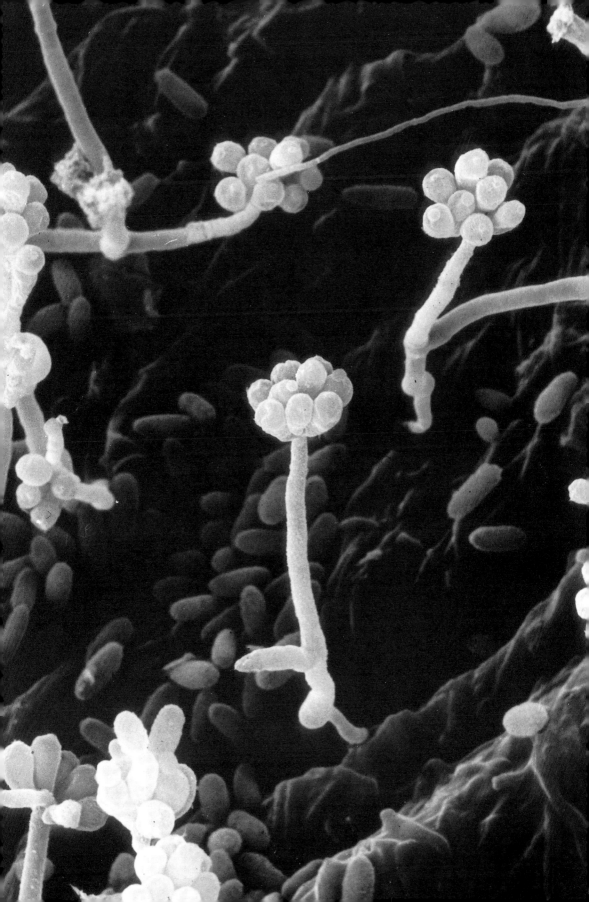

During the Second World War, *S. chartarum* effectively spoiled sandbags, fibre boards and tents.

Building aspects: Indoors, *Stachybotrys* grows on papered surfaces wetted by water penetration or severe condensation. Heavily wetted gypsum elements used for interior walls and ceilings can be attacked by *Stachybotrys*. This is of particular concern because it produces very toxic substances during growth. This also applies to the agricultural aspects.

Agricultural aspects: Rainfall and high humidity favour contamination with *Stachybotrys* on hay, straw and other cellulose-rich plant material.

Medical aspects: When growing on harvested crops or other plant materials, this fungus produces very toxic metabolites. Contaminated material is not only toxic by ingestion, as is the usual feature of mycotoxins. The fungus can produce substances – macrocyclic trichothecenes – which can irritate the skin and mucosa. Furthermore, they strongly inhibit protein synthesis and the RNA and DNA synthesis. They influence the normal immune response by being immunosuppressive. Inhaled trichothecenes are also toxic.

The best described toxicoses are from domestic animals caused by ingestion of contaminated hay and straw or from inhalation of infected material used for bedding. The typical clinical signs in horses are divided into three stages, where stage one includes ulceration of the oral mucosa, excessive salivation, rhinitis, conjunctivitis and elevated temperature. Stage two does not exhibit many symptoms, but the blood will have a reduced number of white blood cells and platelet cells. Stage three leads to complete inability of the blood to clot, elevated temperature, diarrhoea and secondary bacterial infections. An atypical shock form may develop 10-12 hours following ingestion of very toxic material. Nervous signs are the dominating symptoms. Spasms and tremors are present before death.

Human affection is also described. Dermatitis, a burning sensation in the mouth and nasal passage and mucosal irritation such as cough, phlegm and itching of nose, throat and eyes are experienced. Inhalation of conidia may also induce pathological changes (pneumomycotoxicoses).

Specific metabolites

Several toxic secondary metabolites have been isolated and characterised from *S. chartarum*. They include the trichothecenes verrucarin J, roridin E and satratoxin F, G and H, and are all responsible for the dermatotoxic and cytotoxic symptoms described above. All the components are chemically stable, which explains the fact that contaminated hay keeps its toxic properties for many years.

Satratoxin H

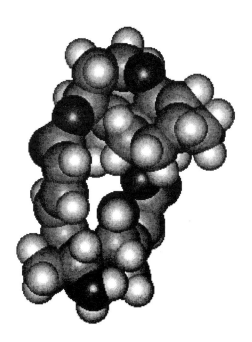

Thamnidium elegans

From Greek *thamnos*: a bush, *eidos*: like;
from Latin *elegans*: elegant.

zones. It prefers low temperatures (psy-chrophilic) and adapts well to temper-atures of 1-2°C where good growth has been observed. Optimal growth is at 18°C, maximum tolerated is 27°C.

It can be found on different kinds of dung (i.e., it is coprophilous) and is isolated from cereals, cold-stored meat and from house dust and air.

Practical application

This mould has so far no practical appli-cation but is included because it is com-mon and spectacular.

Damaging effects

No special damaging effects are report-ed.

Specific metabolites

Relatively large amounts of ethylene are produced from cultures on corn-steep-glucose medium.

Description

The characteristic colonies of this ele-gant bush-like fungus are easily recog-nised and will reach 5 cm in diameter on different media in 7 days.

Colonies are 1.5-2 cm high. A typical sporangiophore has one large sporangi-um which often reaches the lid of the Petri dish along with several shining white dichotomously branched spor-angioles with 2-6 spores. Both types of spores are oval and of similar size: 6.5-15.5 × 4.5.-7.5 μm.

Ecology

T. elegans is very common in temperate

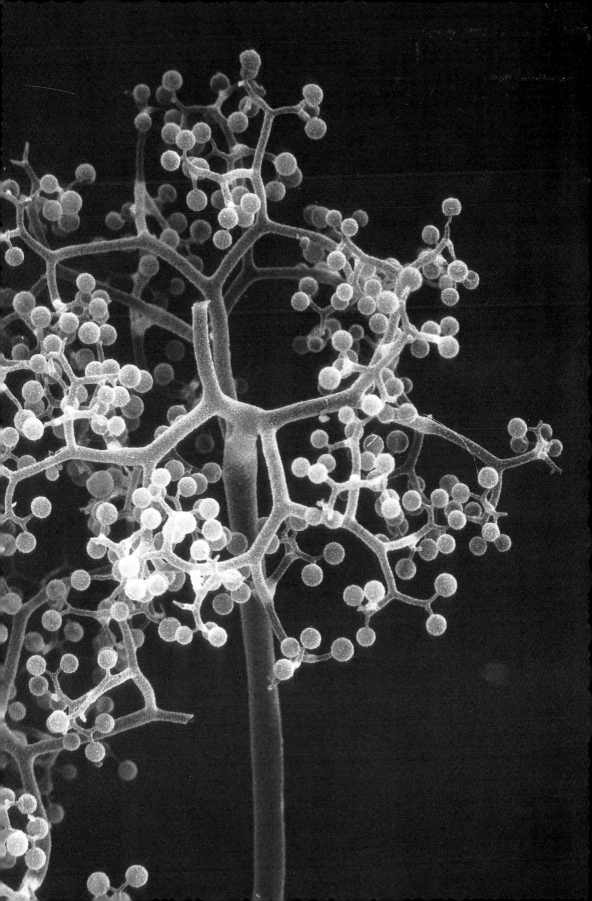

Trichoderma viride

From Greek *thrix*: hair, *derma*: skin;
from Latin *viridis*: green.

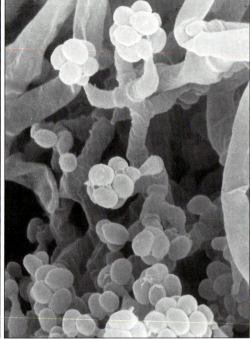

Description

This fungus has a characteristic effuse growth with whitish to bright green colonies reaching a diameter of around 7 cm in 5 days at 25°C on V-8.

The conidia are almost globose with a distinctly rough surface, 3.6-4.5 μm in diameter. In culture many strains of this species have a distinct smell of coconut.

Ecology

Trichoderma is one of the most widespread soil fungi, isolated from numerous different geographical locations. Because of the strong cellulolytic activity, it is an early coloniser of the litter layer. It is found on fallen timber and driftwood and in polluted streams, sewage treatment plants, composted household refuse and fields treated with activated sludge. In damp houses it is found on paper, on tapestry and in kitchens on the outer surface of unglazed ceramics.

The species is often isolated from indoor air samples and is present in house dust from most parts of the world.

Temperature for good growth ranges between 6 and 32°C with optimum temperature from 20-28°C. Some isolates are able to grow at 37°C.

Trichoderma viride is not very sensitive to antibiotics produced by other micro-organisms, but growth is inhibited by tannins from leaf-litter, by some bacterial volatile compounds and by presence of the mould *Stachybotrys chartarum*.

Practical application

Cellulase and hemicellulase are produced commercially by this mould for use in production of beer, wine and food processing, such as enhancing the aroma in tea and mushroom products.

The cellulolytic activity of *T. viride* is also utilised for biochemical testing of material containing cellulose in the field of biodeterioration research.

When tested for antibiotic activity, *T. viride* appears to possess such characteristics, but unfortunately the substances are too toxic for mammals, including humans.

Fungal pellets of another species of *Trichoderma* (*T. harzianum*) have been used mixed with ground bark to protect different trees and vegetable crops against infections from pathogens.

Damaging effects

T. viride can be the cause of storage rot in grain, nuts, citrus fruits, onions and tomatoes. Water supplies may occasionally be affected by this fungus.

Building aspects: Building material damaged by water may be affected by heavy growth of *T. viride*, e.g. wood constructions in roof or floor, or mineral fibre panels used for insulation soaked by water penetrating the panels, with subsequent loss of resistance to micro-organisms.

Medical aspects: Type 1 allergy is described, but is relatively rare. Inhalation of conidia or the volatile components of this mould may cause symptoms as described under *Stachybotrys chartarum*, as some of the metabolic substances are chemically closely related to trichothecenes.

Specific metabolites

The characteristic smell of coconut from cultures of *Trichoderma viride* is derived from the metabolite 6-pentyl-α-pyrone, which is a ketone. Other metabolites of *Trichoderma* are pachybasin, chrysophanol and emodin. Other substances are trichodermin, related to trichothecin, mentioned earlier and known to be very toxic, and trichotoxin A.

Substances with antibacterial activity are suzukacillin and dermadin. A volatile antibiotic component active against different moulds has been isolated.

T. harzianum has been reported to produce the antifungal compounds trichoriazines.

Ulocladium chartarum

From Latin *charta*: paper.

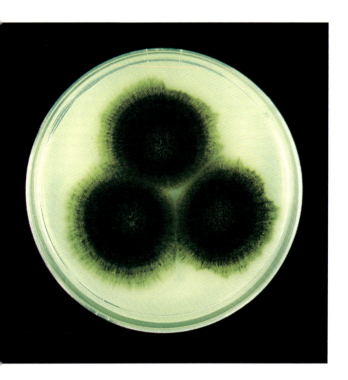

Description

To the untrained eye, *Ulocladium* has a striking resemblance to *Alternaria alternata*. Like this fungus, colonies are black to olivaceous black, reaching a diameter of 5.5 cm on MEA in 7 days when grown at 25°C.

The conidiophores are golden brown, often geniculate with pores.

Conidia are either solitary or in chains, often short ellipsoidal with transverse and longitudinal septa 1-5 in number.

The colour can be brown to olivaceous, the surface smooth to verrucose. Solitary or in chains 18-38 × 11-20 μm in size.

At first the base of the conidium is conical, rounded with age. Apex broadly rounded before production of a so-called "false beak" which functions as a conidiophore for the formation of secondary conidia. This makes the superficial resemblance with *Alternaria* conidia. These have, however, gradually tapering true beaks.

Ecology

Ulocladium frequently occurs in air and dust samples. It is isolated from soil, wood, decaying plant material,

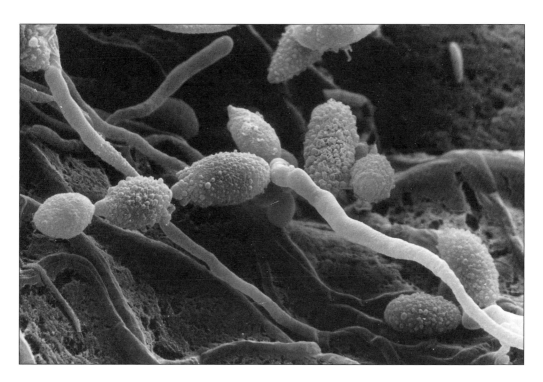

paper and textiles. Temperature for growth ranges from 5°C to 34°C.

Practical application

None.

Damaging effects

Under humid conditions, paper, tapestry and paint can be attacked and spoiled by *Ulocladium chartarum*. Also found on water-damaged building materials, e.g. gypsum boards.

Medical aspects: Positive skin prick test to *U. chartarum* is seen in patients with airway allergy to *Alternaria*. This is due to the fact that the most important allergen in *Ulocladium* and *Alternaria* (Alt a 1) has more or less the same protein structure. Among the fungi, it is unusual to find this partial identity at an antigenic level between different genera.

The clinical consequence is that *Ulocladium* contributes to the allergenic burden of *Alternaria*-sensitive patients.

Metabolites

None known.

Wallemia sebi

From Latin *sebi*: tallow.
Synonyms: *Sporendonema sebi, S. epizoum, Torula pulvinata, Hemispora stellata, W. ichtyophaga.*

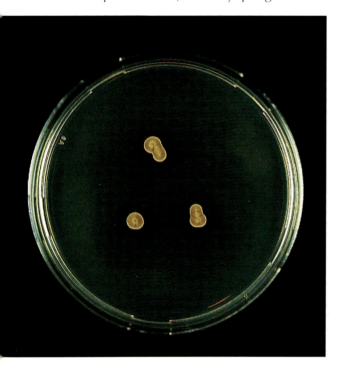

Description

The very slow growing colonies will reach a diameter of about 0.5 cm in 10 days when grown on MEA with 20% or 40% sucrose added. The colonies are fan-like or stellate, usually powdery, orange-brown to blackish brown.

The conidiophores vary in length; they are cylindrical, smooth-walled, chocolate to pale brown and end in fertile hyphae which separate and fall apart as 4 conidia. Initially they are cubic and become more or less globose, 2.5-3.5 μm in diameter. Conidia are pale brown with a finely warted structure of the wall.

Ecology

This neglected mould is much more common than reported. However, due to the xerophilic properties (prefers growth under dry conditions), it will be overgrown or not observed on conventional media. But given the proper growing conditions with low water activity through addition of sugar or salt in high concentrations, it will exhibit a worldwide distribution and is found in air samples, house dust, soil and dry foodstuffs such as jam, marzipan, dates, bread, bacon and salted fish. It also occurs on fruit, hay and textiles.

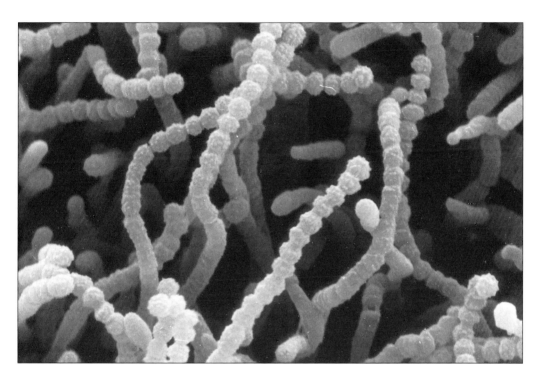

Optimal temperature for growth is 23-28°C with maximum at 36°C.

Practical application

None.

Damaging effects

W. sebi may attack foodstuffs with low water activity (available water) such as salted fish or frosted cake. In milk products such as sweetened condensed milk, button-like pellets of *W. sebi* can be found. Tobacco and fatty materials can also be attacked.

From Japan, allergy to *W. sebi* is reported. In a Danish group of allergic patients, nobody responded positively to skin prick test with this mould.

Specific metabolites

Walleminol, a mycotoxin, is the only one verified until now. However, tryptophol and UCA 1064-ß have also been identified from *W. sebi*. The production of "wallemia A-E" has not yet been confirmed.

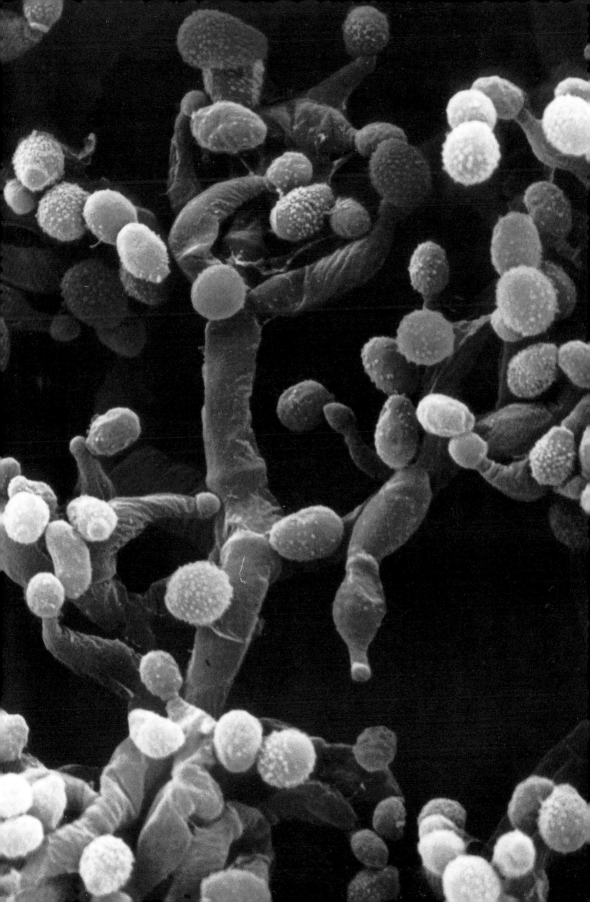

Photographs

Abbreviations

The photographs are from the authors except:
Ole Filtenborg: cover
Joel Andersen Fig. 6
Peter A. Nielsen Fig. 13
John Ouellette Fig. 14
Dansk Bygningsanalyse Fig. 18
Hans P. van Egmond Fig. 23
Susanne Juhl Andersen Fig. 30
José Bresciani & Jørgen Grønvold Fig. 35 a + b
Harry C. Evans Fig. 40
Bent Petersen Fig. 52
Bo Holma Fig. 53
Susanne H. Sparholt Fig. 54
Per Stahl Skov Fig. 55
Brian Flannigan Fig. 57
Bente Schwartz Fig. 58
Jens T. Mortensen Fig. 60
Erik Foged Fig. 65
Department of Dermatology, University Hospital,
    Copenhagen Fig. 69
Preben Løvgreen Nielsen Fig. 70
Aksel Stenderup Fig. 71, Fig. p. 98

APC:	antigen-presenting cell
CIE:	crossed immunoelectrophoresis
CREA:	creatine sucrose agar
CRIE:	crossed radioimmunoelectrophoresis
CYA:	Czapek yeast autolysate agar
CY20S:	Czapek yeast extract 20% sucrose agar
CZA:	Czapek agar
DG18:	Dichloran 18% glycerol agar
DRBC:	dichloran rose bengal chloramphenicol agar
DRYES:	Dichloran rose bengal yeast extract sucrosa agar
GRAS:	generally regarded as safe
IgE:	immunoglobulin E
IgG:	immunoglobulin G
MEA:	malt extract agar (Blakeslee formula)
ODTS:	organic dust toxic syndrome
PSA:	Potato sucrose agar
SAB:	Sabauraud dextrose agar
YES:	yeast extract sucrose agar

Bibliography

General mycology

Kendrick, B. The fifth kingdom. 2nd ed. Mycologue Publications/ Focus Texts, Newburyport. 1992. 406 pp.

Ingold, C.T. and H.J. Hudson: The biology of fungi. 6th ed. Chapman & Hall, London. 1993. 224 pp.

Gams, W., H.A. van der Aa, A.J. van der Plaats-Niterink, R.A. Samson and J.A. Stalpers: CBS course of mycology. 3rd ed. Centraalbureau voor Schimmelcultures, Baarn. 1987. 136 pp.

Domsch, K.H., W. Gams & T.-H. Anderson: Compendium on soil fungi. 2nd ed. IHW-Verlag, Eching. 1993. Two volumes. 859 + 391 pp.

Biodeterioration and food mycology

Samson, R.A. and E.S. van Reenen-Hoekstra (eds.): Introduction to food-borne fungi. 3rd edition. Centraalbureau voor Schimmelcultures, Baarn. 1988. 299 pp.

Pitt, J.I. and A.D. Hocking: Fungi and food spoilage. Academic Press, Sydney. 1985. 413 pp.

Arora, D.K., K.G. Mukerji and E.H. Marth (eds.): Foods and feeds. Handbook of applied mycology. Vol. 3. Marcel Dekker, New York. 1991. 621 pp.

Samson, R.A. and J.I. Pitt (eds.): Modern methods in food mycology. Elsevier, Amsterdam. 1992. 388 pp.

Ecology

Carroll, G.C. and D.T. Wicklow (eds.): The fungal community. Its organization and role in the ecosystem. 2nd ed. Marcel Dekker, New York. 1992. 976 pp.

Arora, D.K., B. Rai, K.G. Mukerji and G.R. Knudsen (eds.): Soil and plants. Handbook of applied mycology. Vol. 1. Marcel Dekker, New York. 1991. 720 pp.

Bhatnagar, D., E.B. Lillehoj and D.K. Arora (eds.): Mycotoxins in ecological systems. Marcel Dekker, New York. Vol. 5. 1992. 443 pp.

Biotechnology

Arora, D.K., R.P. Elander, and K.G. Mukerji (eds.): Fungal biotechnology. Handbook of applied mycology vol. 4. Marcel Dekker, New York. 1991. 1114 pp.

Finkenstein, D.B. and C. Bell (eds.): Biotechnology of filamentous fungi. Butterworth-Heinemann, Boston. 1991. 520 pp.

Mycotoxins & mycotoxicosis

Smith, J. and Moss, M.O.: Mycotoxin. Formation, analyses and significance. John Wiley & Son. 1985. 148 pp.

Betina, V.: Mycotoxins. Chemical, biological and environmental aspects. Elsevier, Amsterdam. 1989. 438 pp.

Betina, V. (ed.): Chromatography of mycotoxins. Techniques and applications. Elsevier, Amsterdam. 1993. 436 pp.

Sharma, R.P. and D.K. Salunkhe (eds.): Mycotoxins and phytoalexins. CRC Press, Boca Raton. 1991. 775 pp.

Allergy & other adverse health effects

Mygind, N.: Essential allergy. Blackwell Scientific Publications, Oxford. 1986. 480 pp.

Middelton, E. Jr., Reed, C.E., Ellis, E.F., Franklin Adkinson, N. Jr., Yunginger, J.W. & Busse, W.W. (eds.): Allergy. Principles and practice. Vol. I and Vol. II. Fourth edition. Mosby, St. Louis. 1993. 962 pp. + 831 pp.

Indoor climate and air quality problems. Investigation and remedy. SBI Report 212. Danish Building Research Institute. 1990. 42 pp.

Wanner, H.U., Verhoeff, A., Colombi, A., Flannigan, B., Gravesen, S., Mouilleseaux, A., Nevalainen, A., Papadakis, J. & Seidel K. (eds.): Biological particles in indoor environments. Report No. 12. Commission of the European Communities. 1993. 81 pp.

Mycopathology

Arora, D.K., L. Ajello and K.G. Mukerji (eds.): Humans, animals, and insects. Handbook of applied mycology. Vol. 2. Marcel Dekker, New York. 1991. 783 pp.

McGinnis, M.R.: Laboratory handbook of medical mycology. Academic Press, London. 1980. 661 pp.

Odds, F.C.: *Candida* and candidosis. Second Edition. Baillière Tindall, London. 1988. 480 pp.

Rippon, J.W.: Medical mycology. 3rd ed. W.B. Saunders Company, Philadelphia. 1988. 808 pp.

Taxonomy

Rossman, A.Y., M.E. Palm and L.J. Spielman: A literature guide for the identification of plant pathogenic fungi. APS Press, American Phytopathological Society, St. Paul. 1987. 252 pp.

Barnett J.A., R. W. Payne and D. Yarrow: Yeasts, characteristics and identification. Second edition. Cambridge University Press, Cambridge. 1990. 1002 pp.

Arx, J.A. von: The genera of fungi sporulating in pure culture. 3rd ed. J. Cramer, Verduz. 1981. 424 pp.

Carmichael, J.W., W.B. Kendrick, I.L. Connors and L. Sigler: Genera of hyphomycetes. University of Alberta Press, Edmonton. 1980. 386 pp.

Ellis, M.B.: Dematiaceous hyphomycetes. Commonwealth Mycological Institute, Kew. 1971. 608 pp.

Ellis, M.B.: More dematiaceous hyphomycetes. Commonwealth Mycological Institute, Kew. 1976. 507 pp.

Sutton, B.: The Coelomycetes. Fungi imperfecti with pycnidia, acervuli and stromata. Commonwealth Mycological Institute, Kew. 1980. 696 pp.

Glossary

allergen: antigen which evokes allergy

allergy: hypersensitivity mediated by hyperreactions of immunological mechanisms

anamorph: imperfect (or asexual state) conidial form of sporulation

annellide: a conidiogenous cell which forms blastoconidia in basipetal succession, each conidium is produced through the scar of the previously formed conidium and leaves a ringlike scar (annellation) at the fertile apex after seceding. The conidiogenous cell elongates during conidiogenesis (progressive)

antibody: immunoglobulin, a protein in the blood produced by B-lymphocytes, for example IgE and IgG

antigen: (anti = against, gen = former) antibody former, a component of mainly proteinaceous origin invading the body

ascospores: sexual spore produced in an ascus

ascus: meiosporangium in which the ascospores are formed after karyogamy and meiosis

aseptate: without a crosswall

B-lymphocytes: small white blood cells formed in the bone marrow

basidiomycetous: from a subdivision of fungi in which meiospores (basidiospores) are formed exogenously from a meiosporangium (basidium)

basophil cells: white blood cells circulating in the blood able to bind IgE to its surface

budding: a process of vegetative multiplication in which there is a development of a "daughter" cell from a small outgrowth of a "mother" cell (monopolar, bipolar and multilateral budding, see section of chapter 1 on yeasts)

chlamydospore: a thick-walled, thallic, terminal or intercalary resting propagule (conidium); usually non-deciduous

colony: (of mycelial fungi) a group of hyphae (with or without conidia) which arise from one spore or cell. Colony appearance varies e.g.: velvety, floccose (cottony), funiculose (hyphae aggregated into strands), fasciculate (in little groups or bundles), granulose, powdery, synnematous (compact groups of erect and sometimes fused conidiophores)

columella: an usually swollen sterile central axis within a sporangium, e.g. in Mucorales

complement system: system of 11 proteins in human serum acting in the immune defence

conidiogenesis: process of conidium formation

conidiogenous cell: the fertile cell from which or within conidia are directly produced. Conidiogenous cells may be morphologically identical with or differentiated from vegetative cells (vegetative hyphae)

conidiophore: specialized hypha, simple or branched, on which conidiogenous cells are borne

conidium: asexual, vegetative, nonmotile propagule, not formed by cleavage (as in sporangiospores). In Deuteromycetes (fungi imperfecti = mitotic fungi) the term "conidium" is recommended and the term "spore" is reserved for zoo-, sporangio-, and basidiospores

echinulate: describes cell wall surfaces (e.g. of conidia, or conidiogenous cells), with small pointed processes or spines

exudate: droplets excreted by the mycelium. Can be characteristic for a species e.g. in *Penicillium chrysogenum*

helminthicides: chemical compounds toxic towards worms

herbicides: chemical compounds toxic towards plants (weeds)

histamine: a mediator substance with effects on muscles, blood vessels and mucous glands

holomorph: the whole fungus, including the asexual (imperfect or anamorph) state as well as the sexual (perfect or teleomorph) state

Hülle cells: terminal or intercalary thick-walled cells usually surrounding the ascomata (in some *Aspergillus* species)

hyaline: transparent or nearly so

hypha (pl. hyphae): (vegetative) filament of a mycelium, without or with crosswalls

immune complex: chemical reaction product between antigen and antibody

immunity: resistance, e.g. protection against

infections either during earlier infections or during vaccination

immunoglobulins: antibodies (group of related proteins)

insecticides: chemical compounds toxic towards insects

local treatment: treatment with medicine which has a topical effect; for example, treatment of ringworm with ointment

macroconidia: the larger conidia of a fungus, usually multicelled (e.g. *Fusarium*)

macrophage: (macro = big; phage: eater), big white blood cells which destroy and remove antigens from the body

mast cell: cells in surface tissues which contain histamine granules. Histamine is liberated following contact between antigen and antibodies on the surface of the mast cell. This is part of the mechanism behind asthma and hay fever

merosporangium: (of Mucorales) a cylindrical outgrowth from a swollen end of the sporangiophore, in which asexual merospores are produced in a linear row

metabolite: any substance produced during metabolism

metula (pl. metulae): apical branch(es) of conidiophore-bearing phialides (e.g. *Penicillium*, *Aspergillus*)

microconidia: small conidia, usually one-celled

mycelium: (vegetative) mass of hyphae; thallus (vegetative) body of the fungus

mycotoxin: a metabolite produced by a fungus that is toxic towards vertebrates

oblate: formed as a flat sphere

phialide: a conidiogenous cell which produces conidia (phialoconidia) in basipetal succession, without an increase in length of the phialide itself

pseudo-mycelium: the formation of a filamentous structure consisting of cells which arise exclusively by budding

sclerotium: a resting body, usually globose, produced by aggregation of hyphae into a firm mass with or without host tissue, normally without conidia (sterile)

septum: a crosswall in a cell

specific metabolite = secondary metabolite = natural product: differentiation products in living organisms being produced only by some species in a family

spine: a narrow process with a sharp point

spinulose: delicately spiny

sporangiole: small, usually globose sporangium with a reduced columella and containing one or a few spores

sporangiophore: a specialized hyphal branch which supports one or more sporangia

sporangiospore: a spore produced within a sporangium

sporangium: asexual reproductive structure, unicellular, in which spores are produced by cytoplasmic cleavage

spore: a general term for a reproductive structure in fungi, bacteria, and cryptogams. In fungi the term "spore" is used in several combinations e.g. chlamydospore, ascospore, zoospore, basidiospore

sporodochium: a cushion-like mass of conidiophores, conidia and conidiogenous cells produced above the substrate

stipe: stalk

stolon: a "runner" (e.g. *Rhizopus*)

striate: marked with lines, grooves or ridges

stroma: a mass or matrix of vegetative hyphae, with or without tissue of the substrate, in or on which fructifications can be produced

substrate: the material on or in which an organism is living

synnema (pl. synnemata): bundles of erect hyphae and occasionally fused conidiogenous cells bearing conidia (also called coremia)

systemic treatment: treatment with medicine with effect in the whole body

T-lymphocytes: small white blood cells, which are formed in the bone marrow. They control production of antibodies and produce lymphokines, battle substances

teleomorph: perfect (sexual) state of a fungus; ascigerous or basidial form of sporulation

uniseriate: in one series. In *Aspergillus*, phialides arise directly from the vesicle

velutinous: velvety

verrucose: having small rounded processes (warts)

verruculose: delicately verrucose

verticil: whorl

vesicle: a bladder-like sac; the swollen apex of the conidiophore in *Aspergillus*

xerophilic: preferring dry places, growing under dry conditions

zygospore: a sexual spore produced by Zygomycetes, thick-walled, often ornamented and darkly pigmented, formed by fusion of a pair of gametangia

Descriptions of some common microfungi: illustrations

Page	Fungus	Medium	Temperature	Magnification	Comments
73	*Cunninghamella echinulata*			× 4200	SEM
74	*Alternaria alternata*	V-8 agar	25°C		
75	*Alternaria alternata*			× 6000	SEM
77	*Alternaria alternata*			× 4200	SEM
78	*Aspergillus flavus*	CZA	25°C		SEM
79	*Aspergillus flavus*			× 1800	
80	*Aspergillus fumigatus*	CYA	25°C		
80	*Aspergillus fumigatus*			× 2400	SEM
81	*Aspergillus fumigatus*			× 100	macrophotograph
83	*Aspergillus fumigatus*			× 3400	SEM
84	*Aspergillus niger*	CYA	25°C		
84	*Aspergillus niger*			× 700	SEM
86	*Aspergillus ochraceus*	CYA	37°C		
87	*Aspergillus ochraceus*			× 500	SEM
88	*Aspergillus oryzae*	CZA	25°C		
89	*Aspergillus oryzae*			× 1000	SEM
90	*Aspergillus terreus*	CYA	37°C		
91	*Aspergillus terreus*			× 5000	SEM
92	*Aspergillus versicolor*	CYA	25°C		
93	*Aspergillus versicolor*			× 2300	SEM
94	*Aureobasidium pullulans*	V-8 agar	25°C		
94	*Aureobasidium pullulans*			× 500	Light micrograph
96	*Botrytis cinerea*	V-8 agar	25°C		
97	*Botrytis cinerea*			× 2300	SEM
98	*Candida albicans*	SAB	37°C		
99	*Candida albicans*			× 4500	SEM
100	*Cladosporium herbarum*	V-8 agar	25°C		
101	*Cladosporium herbarum*			× 8200	SEM
103	*Cladosporium herbarum*			× 4400	SEM
104	*Emericella nidulans*	CYA	25°C		
104	*Emericella nidulans*			× 4500	ascomata
105	*Emericella nidulans*			× 1000	anamorph
106	*Epicoccum nigrum*	MEA	25°C		
107	*Epicoccum nigrum*			× 4200	SEM
108	*Eurotium chevalieri*	YES	25°C		
109	*Eurotium chevalieri*			× 4500	anamorph
110	*Eurotium rubrum*	CY20S	25°C		
110	*Eurotium rubrum*			× 4800	SEM
111	*Eurotium rubrum*			× 1300	ascomata
112	*Fusarium graminearum*	PSA	25°C		
113	*Fusarium graminearum*			× 1500	SEM
116	*Fusarium oxysporum*	PSA	25°C		
116	*Fusarium oxysporum*			× 700	SEM
118	*Mucor plumbeus*	V-8 agar	25°C		
118	*Mucor plumbeus*			× 250	SEM
119	*Mucor plumbeus*			× 2200	SEM
121	*Mucor plumbeus*			× 4000	SEM
122	*Paecilomyces variotii*	V-8 agar	25°C		
123	*Paecilomyces variotii*			× 1800	SEM

124	*Penicillium aurantiogriseum*	YES	25°C		
125	*Penicillium aurantiogriseum*			× 2800	SEM
126	*Penicillium camemberti*	CYA	25°C		
126	*Penicillium camemberti*			× 2000	SEM
127	*Penicillium camemberti*			× 2300	SEM
128	*Penicillium chrysogenum*	CYA	25°C		
129	*Penicillium chrysogenum*			× 2400	SEM
130	*Penicillium expansum*	CYA	25°C		
131	*Penicillium expansum*			× 3300	SEM
132	*Penicillium glabrum*	CYA	25°C		
132	*Penicillium glabrum*			× 1000	SEM
133	*Penicillium glabrum*			× 4200	SEM
134	*Penicillium polonicum*	YES	25°C		
135	*Penicillium polonicum*			× 2000	SEM
136	*Penicillium roqueforti*	CYA	25°C		
137	*Penicillium roqueforti*			× 1000	SEM
138	*Rhizopus stolonifer*	V-8 agar	25°C		
139	*Rhizopus stolonifer*			× 600	SEM
140	*Scopulariopsis brevicaulis*	V-8 agar	25°C		
141	*Scopulariopsis brevicaulis*			× 2200	SEM
143	*Scopulariopsis brevicaulis*			× 3800	SEM
144	*Stachybotrys chartarum*	V-8 agar	25°C		
145	*Stachybotrys chartarum*			× 2000	SEM
148	*Thamnidium elegans*	V-8 agar	25°C		
149	*Thamnidium elegans*			× 2000	SEM
150	*Trichoderma viride*			× 2200	SEM
152	*Ulocladium chartarum*	V-8 agar	25°C		
153	*Ulocladium chartarum*			× 1500	SEM
154	*Wallemia sebi*	DG 18	25°C		
155	*Wallemia sebi*			× 2000	SEM
157	*Trichoderma viride*			× 5400	SEM

Index

166